Nanosystems Design and Technology

Giovanni De Micheli • Yusuf Leblebici
Martin Gijs • Janos Vörös

Nanosystems Design
and Technology

Springer

Prof. Dr. Giovanni De Micheli
EPFL - Swiss Federal Institute
of Technology
Integrated Systems Lab
1015 Lausanne, Switzerland
giovanni.demicheli@epfl.ch

Prof. Dr. Yusuf Leblebici
EPFL - Swiss Federal Institute
of Technology
Microelectronic Systems Lab
1015 Lausanne, Switzerland
Yusuf.Leblebici@epfl.ch

Prof. Dr. Martin Gijs
EPFL - Swiss Federal Institute
of Technology
Microsystems Lab
1015 Lausanne, Switzerland
martin.gijs@epfl.ch

Prof. Dr. Janos Vörös
ETHZ - Swiss Federal Institute
of Technology
Inst. Biomedizinische Technik
Gloriastrasse 35 - 8092
Zürich, Switzerland
janos.voros@biomed.ee.ethz.ch

ISBN 978-1-4419-0254-2   e-ISBN 978-1-4419-0255-9
DOI 10.1007/978-1-4419-0255-9
Springer Dordrecht Heidelberg London New York

Library of Congress Control Number: 2009926034

Printed on acid-free paper

Springer is part of Springer Science+Business Media (www.springer.com)

Foreword

It is a great pleasure to write for this book, which largely presents research funded by the Competence Centre for Materials Science (CCMX) of the Swiss Federal Institutes of Technology (ETH domain). CCMX was created in early 2006 with the aim of networking research groups within the ETH domain and interfacing to the research needs of Swiss industry. Materials science and technology is a transversal subject that addresses the diverse materials used in many different industrial sectors; thus CCMX focuses on several sectors of present and future economic importance including Materials for Micro- and Nanosystems (MMNS) with Giovanni De Micheli as director.

CCMX has provided a new funding model with several research teams working together on interdisciplinary projects. MMNS took up this model of *cooperative engineering* to initiate three large projects, namely: (a) Materials, Devices, and Design Technologies for Nanoelectronic Systems Beyond Ultimately Scaled CMOS, (b) Laboratory on Chip, and (c) Development and Characterization of Nanowires for Applications in Bioelectronics. In these projects several pre- and postdoctoral researchers worked within different teams with regular meetings for an exchange of ideas and technology. Partners included the Swiss Federal Institutes of Technology of Lausanne (EPFL) and Zurich (ETHZ), the Swiss Center for Electronics and Microsystems and Microtechnology (CSEM), and the Paul Sherrer Institute (PSI).

Judging by the important and varied results generated in less than 3 years of the project, this way of working has proved highly successful, and this publication provides an excellent opportunity to bring the results to the attention of a wider international audience. We hope that this successful start will be continued in the *nano-tera.ch* project, which has taken over this area, while CCMX continues to fund initiatives in other areas (e.g., metals, powders and coatings, and materials for life sciences). We are grateful to the board of the Swiss Institutes of Technologies (CEPF/ETHrat) for providing this opportunity to demonstrate the effectiveness of cross-team interdisciplinary research.

CCMX
Competence Centre for
Materials Science and Technology

Karen Schrivener
Professor and Director, CCMX

Acknowledgments

The authors would like to acknowledge the contributions of all researchers on the CCMX/MMNS teams that produced the research and results reported here. The authors are indebted to the Board of the Swiss Federal Institutes of Technologies (CEPF/ETHrat) for funding this research through the Competence Center for Materials (CCMX). Additional funding was provided by the Swiss National Science Foundation, the Swiss Federal Office for Professional Education and Technology (OFFT), and the European project MOBESENS and GoodFood.

The authors are also grateful to Stefano Pietrocola and Marta Lombardini for assistance with the experiments reported in Chap. 4, to the technical staff members at EPFL, ETHZ, CSEM, and PSI, with special mention to Martin Lanz for his support with microfabrication. The authors would also like to thank the Applied Molecular Receptors Group (CSIC, Barcelona, Spain) and Unisensor SA (Wandre, Belgium) for providing the immunoassay reagents, Perstorp Specialty Chemicals (Sweden) for providing the hyperbranched polymers, and Arrayon Biotechnology for the biofunctionalization of the WIOS chips.

Last but not least, the authors would like to express their sincere appreciation to Anil Leblebici, who as CCMX coordinator enabled the collaborative research by supporting the infrastructure and in particular coordinated the efforts for producing this book.

Contents

Contributors

Janko Auerswald CSEM – Swiss Center for Electronics and Microtechnology Inc., Neuchâtel, Switzerland

Vaida Auzelyte PSI – Paul Scherrer Institute, Villigen, Switzerland

Stefan Berchtold CSEM – Swiss Center for Electronics and Microtechnology Inc., Neuchâtel, Switzerland

Didier Bouvet EPFL – Swiss Federal Institute of Technology, Lausanne, Switzerland

Edoardo Charbon EPFL – Swiss Federal Institute of Technology, Lausanne, Switzerland

Giovanni De Micheli EPFL – Swiss Federal Institute of Technology, Lausanne, Switzerland

Jean-Marc Diserens Nestlé Research Center, Nestec Ltd., Lausanne, Switzerland

Emile P. Dupont EPFL – Swiss Federal Institute of Technology, Lausanne, Switzerland

László Forró EPFL – Swiss Federal Institute of Technology, Lausanne, Switzerland

Martin A. M. Gijs EPFL - Swiss Federal Institute of Technology, Lausanne, Switzerland

Adrian M. Ionescu EPFL – Swiss Federal Institute of Technology, Lausanne, Switzerland

Haykel Ben Jamaa EPFL – Swiss Federal Institute of Technology, Lausanne, Switzerland

Young-Hyun Jin EPFL – Swiss Federal Institute of Technology, Lausanne, Switzerland

Bahman Kheradmand Boroujeni EPFL – Swiss Federal Institute of Technology, Lausanne, Switzerland
CSEM – Swiss Center for Electronics and Microtechnology, Neuchâtel, Switzerland

Helmut F. Knapp CSEM – Swiss Center for Electronics and Microtechnology Inc., Neuchâtel, Switzerland

Estelle Labonne EPFL – Swiss Federal Institute of Technology, Lausanne, Switzerland

Yusuf Lebelebici EPFL – Swiss Federal Institute of Technology, Lausanne, Switzerland

Ulrike Lehmann EPFL – Swiss Federal Institute of Technology, Lausanne, Switzerland

Yves Leterrier EPFL – Swiss Federal Institute of Technology, Lausanne, Switzerland

Robert MacKenzie ETHZ – Swiss Federal Institute of Technology, Zürich, Switzerland

Arnaud Magrez EPFL – Swiss Federal Institute of Technology, Lausanne, Switzerland

Jan-Anders E. Månson EPFL – Swiss Federal Institute of Technology, Lausanne, Switzerland

Evgeny Milyutin EPFL – Swiss Federal Institute of Technology, Lausanne, Switzerland

Kirsten E. Moselund EPFL – Swiss Federal Institute of Technology, Lausanne, Switzerland

Paul Muralt EPFL – Swiss Federal Institute of Technology, Lausanne, Switzerland

Cristiano Niclass EPFL – Swiss Federal Institute of Technology, Lausanne, Switzerland

Sven Olliges ETHZ – Swiss Federal Institute of Technology, Zürich, Switzerland

Christian Piguet CSEM – Swiss Center for Electronics and Microtechnology Inc., Neuchâtel, Switzerland

Giovanni A. Salvatore EPFL – Swiss Federal Institute of Technology, Lausanne, Switzerland

Alexandre Schmid EPFL – Swiss Federal Institute of Technology, Lausanne, Switzerland

Maximilian Sergio EPFL – Swiss Federal Institute of Technology, Lausanne, Switzerland

Nava Setter EPFL – Swiss Federal Institute of Technology, Lausanne, Switzerland

Harun H. Solak PSI – Paul Scherrer Institute, Villigen, Switzerland

Ralph Spolenak ETHZ – Swiss Federal Institute of Technology, Zürich, Switzerland

Milos Stanisavljevic EPFL – Swiss Federal Institute of Technology, Lausanne, Switzerland

Igor Stolitchnov EPFL – Swiss Federal Institute of Technology, Lausanne, Switzerland

Guillaume Suárez CSEM – Swiss Center for Electronics and Microtechnology Inc., Neuchâtel, Switzerland

Guy Voirin CSEM – Swiss Center for Electronics and Microtechnology Inc., Neuchâtel, Switzerland

Janos Vörös ETHZ - Swiss Federal Institute of Technology, Zürich, Switzerland

Chapter 1
Nanosystems

Giovanni De Micheli

Introduction

This book addresses the impact of nanotechnologies on the design of electronic systems. We use the term "system" in the broad sense, and thus we consider technologies that range from nanoelectronics to sensing and micro-/nanofluidics and that encompass techniques bridging the gap between engineering and biology. We also use the term *nanosystem* to indicate that the system embodies nanodevices. As the application space of nanosystems is very vast, we will focus on specific topics that exemplify the challenges and opportunities that nanotechnology brings to the system design table.

This book also stresses the interdisciplinarity of research in this field. New (nano) materials are key constituents of circuits and sensors, which in turn support the design of specific integrated structures, commonly called architectures (or nanoarchitectures). Electronic systems may be realized in different forms such as systems on chips (SoCs), systems in package (SiPs) or 3-dimensional (3D) integrated systems, and they can provide services by operating either in isolation or within a network. The use of nanodevices in integrated services is important for achieving specific technical goals and motivates the interest of the system industry in new materials. Nevertheless, whereas new materials are key to system innovation, profits derive mainly from the sales of services enabled by nanosystems. As an example, new nanostructured surfaces allow us to design and manufacture biosensors, which are components of labs-on-chips (LoCs). These LoCs can be used at medical points of care, with a significant reduction in health maintenance costs. Whereas in general the market potential for systems and services is large, the competitive market pressure on electronic components and derivatives forces hardware companies to sell them with small profit margins, except for specific niche areas.

At the same time, services and system requirements motivate the design of new hardware platforms, with new architectures and circuits. The stringent requirements on circuits and sensors motivate the use of new materials and the introduction

G. De Micheli
EPFL – Swiss Federal Institute of Technology, Lausanne, Switzerland

G. De Micheli et al., *Nanosystems Design and Technology*,
DOI 10.1007/978-1-4419-0255-9_1, © Springer Science+Business Media LLC 2009

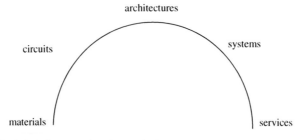

architectures

circuits

systems

materials

services

Fig. 1.1 The food chain in system/service products

of these new materials within standard semiconductor manufacturing processes. Examples are the addition of gold and thiols on top of standard CMOS processes to connect DNA probes in the case of intelligent integrated biosensors. Thus we can take the viewpoint that nanomaterials research is driven by the system/service requirements. Indeed the results described in this book are motivated by the objectives of satisfying systems needs, such as finding alternative solutions to computational systems beyond ultimately scaled CMOS technology, searching for materials and sensor support for LoCs dedicated to diagnostics, food analysis, and environmental analysis, and looking at advanced fabrication technologies for nanostructures. It is also a firm belief that scientific and technical innovation will stem from research at the intersection among disciplines, where creativity and ingenuity can lead to major steps forward (Fig. 1.1).

It is also interesting to address the potentials and limitations of electronic systems with specific reference to their enabling technologies. This analysis will help us to understand what technologies will be needed to progress further. When referring to "technologies" we use this word in a broad sense as well. It encompasses manufacturing and design technologies (DTs). The former represent ways of processing materials and represent an engineering view of nanotechnology. The latter are the enabling methods to carry complex integrated projects to completion. From a practical standpoint, software means are ubiquitously used for the conception, design, and runtime operation of most electronic products. Software design tools, also called computer-aided design (CAD) tools and methods, have been instrumental in the growth and success of electronic products [12]. Broadly speaking, we will refer to this body of knowledge as DTs. Such "soft" technologies, together with the "hard" manufacturing technologies (MTs), are tightly interrelated.

The Nano Landscape

The scientific, technological, and economic landscape of nanotechnology spans several axes. On one side, the market economy demands better products, where better is often equated with faster, smaller, and cheaper. This demand is often satisfied by the unprecedented offers of various technologies, which have several desirable characteristics, such as the small size of nanodevices, their corresponding fast switching speed, their dimensional match with biological compounds, and the potentials for ultra-large-scale integration. Even though the practical use of some nanotechnologies has not yet been assessed, their potential use sustains R&D for many advanced products. In this perspective, it is important to review the overall technological constraints on SoC and LoC architectures, to assess the potential benefit of new materials and the corresponding circuits and architectures.

The Social and Economic Pull

We face several challenges where technology can be instrumental in improving the quality of living. The following examples are drivers that pull the marketplace. The global economy and society demands that each human be reachable: language barriers are still an impediment for the largest part of the world population. Real-time natural language translation, achieved by portable devices, is a major objective for the engineering community to achieve. Many years of research in this domain [2,10] have shown that the problem has solutions, but the required computational effort is high. This objective represents an important driver for multiprocessor architectures and related MTs. Possible solutions may stem from the combination of nanotechnology with parallel architectures.

An increasing number of untethered devices are being used where energy supply is a critical resource. First we pay a price – in terms of nonrenewable energy consumption – for each service provided to us from electronic products. Next we need to charge, replace, and dispose of batteries. Last but not least, it is inconvenient, and sometimes impossible, to refurnish energy (e.g., change batteries) in some circumstances. Energy-efficient designs, with minimum energy consumption, have been a goal of researchers for over two decades [34]. Today we are approaching the design of systems that can *harvest* energy [31] from the environment, enough to be independent from energy sources. To achieve self-sustainable computation, we need a global rethinking of how electronic devices are designed and how systems are controlled and operated.

The use of LoCs for biodiscovery, health management, and environmental monitoring [30] is also an important goal with a large social value added. The integration of sensing in SoCs and the use of suitable nanomaterials and nanotechnology has opened the door to an unprecedented wave of innovation. Specifically, LoCs can be instrumental for better diagnosis and to achieve more affordable medical care.

In summary, the economic and social pull of electronics is strong: we will need increasingly larger processing power and efficiency as well as new means to interact with the environment and to communicate. There is no reason to believe that we will be satisfied with computing and communication systems as they are now; indeed nanotechnologies present us with untapped opportunities to address market and social problems.

The Technology Push

At of this writing, CMOS circuits designed with a 45-nm node are in production and use. There is strong reason to believe that products designed with the 32/22-nm CMOS nodes will be realized within a decade [20]. Still, technical and economic difficulties plague the technology growth. Since the setup of manufacturing in each new technology node requires large capital expenditures, few companies – today linked mainly by technology alliances – can afford advanced design.

Nanotechnologies have a high potential to improve electronic circuits and systems. Before describing technological details, we need to realize that the term "nano" is used in different contexts. For example, the current 45-nm CMOS processes can be considered nanotechnologies, even though these processes are an evolution of current MTs. There is a continuum spectrum of solutions between evolutionary and revolutionary technologies. The fact that a technology becomes "disruptive" is related to extrinsic factors such as the ability to significantly lower manufacturing costs, power consumption, and raising performance. Most advanced technology studies in CMOS, beyond the 45-nm node, already use a plethora of materials and plan on using tridimensional transistors, thus departing from the principle of planarity that has characterized silicon technologies. Recent research on silicon nanowires (SiNWs) has shown the possibility of achieving transistors with interesting characteristics, such as abrupt transitions in the I–V plane, which is important for achieving low power dissipation [28]. Moreover, nanowires can be arranged to create integrated computation and wiring structures, thus supporting regular and predictable design methodologies. It is also important to remember that SiNWs mix and match well with CMOS [17] and thus support the realization of specific-purpose macros (Fig. 1.2).

The future of *graphene* and *carbon nanotubes* (CNTs) is promising but still hard to predict. Metallic CNTs perform well as interconnect due to their high thermal and electrical conductivity [9]. Semiconductor CNTs can be used to create transistors, even though an open question is whether or not it is possible to design large-scale circuits with them. Hybrid technologies that use silicon and CNTs are the subject of current investigation, as CNTs provide us with switching devices with higher carrier mobility (as compared to silicon). The current difficulty in realizing long straight sequences of CNTs has prompted design methods and tools to achieve correct and robust design. An example is the choice of specific layout styles and rules to avoid spurious connections [6, 32] (Fig. 1.3).

Molecular switches provide another interesting technology. Some molecules, like *rotaxane*, have two stable states: one conducting and one nonconducting. When these molecules are placed in specific positions, like at the cross-point of a wire array, their state can provide a means for storing information or for performing computation. Indeed, architectures reminiscent of nonvolatile memories [8] or programmable-logic arrays (PLAs) can be designed efficiently with molecular electronics.

The confinement of semiconductor carriers along the three spatial dimensions gives rise to *quantum dots* that can operate as *single-electron transistors* and show

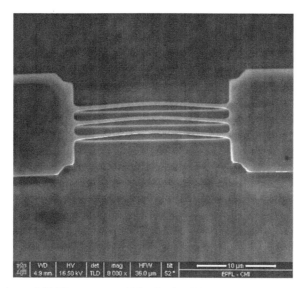

Fig. 1.2 Experimental SiNWs (courtesy of Moselund and Ionescu)

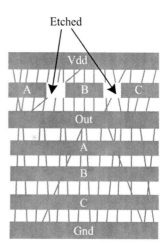

Fig. 1.3 CNFET NAND3 layout with etched regions to be resilient to CNT misalignment

the Coulomb blockade effect. Quantum dots can implement *qubits* for *quantum information processing*. Whereas the potential of this technology is large and disruptive, its practical applications (except for quantum cryptography) are still on the distant horizon. Nevertheless, the interest in the quantum computing paradigm has already spurred research on DT for circuits implementing qubits, e.g., in physical design [25], clocking [44], and synthesis [37].

When facing the prospect of using new MTs for future SoCs, a few important questions come to mind. First of all, are these new technologies suitable (and ready) for system design? We have seen some circuit demonstrators, but designing robust large-scale systems – as is done in CMOS – requires a set of specific characteristics. The large investment in capital and expertise in CMOS leads us to think that it will be difficult for a new technology to replace CMOS. On the other hand, CMOS enhancement by means of new nanotechnologies and nanomaterials is likely to happen. There exist already several examples of hybridization of technologies, such as using nanowires together with CMOS cells [16] or CNTs to provide interconnection on chips [9].

Another important issue is how to design with these materials and technologies. A superficial look at the issue might suggest that it is just a question of changing the back-end of physical design tools. But when looking at the integration of logic and physical design, as well as manufacturing process variability and device dependability issues, it seems appropriate to rethink the entire design flow of nanosystems.

Requirements for Computing Nanoarchitectures

The electronic market is driven by two conflicting goals: achieving *high performance* – as required by multimedia and gaming systems – and achieving *low power* consumption – as needed by all portable systems. In general, both objectives need to be met simultaneously, as high performance systems cannot afford high power consumption and high temperatures (for reliability reasons) and mobile devices need to support complex software applications, thus requiring high-performance processing.

In general, low power consumption is achieved by operating SoCs at low voltage, in combination with voltage (and frequency) scaling and gating. Note that with low-voltage operation (a few tenths of a volt), CMOS transistors will be in weak inversion. The operation at low voltage will impose a limit on the maximum operating frequency and, hence, on the performance delivered by a processor. Thus, parallel operation within multiprocessing is needed to achieve high performance. At the same time, scaling allows us to pack many cores on a single chip [40, 42, 43].

The trend toward multiprocessing SoCs (MPSoCs) is also due to addressing *reliability* issues that may arise from applying deeply-scaled technologies to *life-critical* applications. As scaling may lead to transistors and interconnect with higher *failure rates*, system-level reliability can be insured by *redundancy*, such as having spare processing cores and reconfigurable interconnect means [26].

As a result, there is a strong tendency to compensate the limitations on clock frequency with multiprocessing, as witnessed also by the personal computer market. Whereas MT and DT for multiprocessing are well developed, their efficient use is still limited by the technology of parallelizing software applications (e.g., parallelizing compilers). Moreover, software applications and operating systems need to be considered from the beginning for multiprocessing platforms, and thus multiprocessing does not yet deliver the expected gain on standard applications.

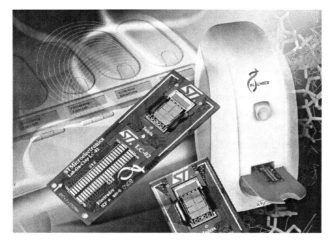

Fig. 1.4 Example of lab-on-a-chip for diagnostics (courtesy of STMicroelectronics)

Requirements for Biomedical Nanoarchitectures

A LoC can be seen as the integration of chemical and biological manipulation on an intelligent substrate [24]. In general, LoCs support microchemistry, and thus can be used for analysis and for synthesis [3] of compounds. In the former domain, LoCs can be used for biodiscovery, environmental monitoring, and medical diagnosis. As for other electronic products, volume production and corresponding competitive cost will be key for acceptance at points of care to enable faster, cheaper, and more precise diagnosis, as well as at other locations, such as mass transport facilities, for effective health control and pandemic prevention. Moreover, future LoCs will be able to support local or distributed computation and access to databases, thus enhancing the effectiveness of diagnosis. This technology can be multifaceted and serve various objectives: it is important for advanced countries where the cost of health care is skyrocketing as well as for developing countries where it is very important to bring medicine at an affordable cost to everyone (Fig. 1.4).

LoCs have many interesting technical features. They show the ultimate hybridization of technologies. Their range of complexity varies. Components that can be integrated in a LoC include, but are not limited to, microfluidics parts for sample transport, sensors to detect proteins/DNA, low-noise electronics, and on-chip data processing algorithms and software to elaborate biological information. LoCs can be programmed (at various levels) to do specific tests. Thus we can envision *field-programmable LoCs* that can be set to do a specific experiment, such as looking for specific compounds in water, according to the circumstances. As in the case of *field-programmable gate arrays* (FPGAs), flexibility, programmability, and volume production reduce the *nonrecurring engineering costs* per unit and are enablers for the broad use of this technology.

Last but not least, LoC technology enables us to research the close relation between nanostructures and living matter. Specific nanostructures, such as nanowires and nanotubes, can be used as supporting material for sensors as well as ways to convey chemicals into precise locations of tissues or cultures. The advantage of nanoscale approaches is that they make it possible to select specific targets, such as specific cells into which to inject chemicals. Interestingly enough, nanostructures can also be constructed using biological materials as scaffold. As an example, DNA strands can combine in predetermined ways and bring together or bond materials tagged to the DNA to make a nanostructure.

Nanosystem Design

The emphasis of this book is on the use of new nanotechnologies for system design. First we address nanoelectronics within advanced SoC design. We consider specific MTs, such as SiNW (for computation), ferroelectric materials (for storage), and piezoelectric materials (for sensors) in Chap. 2. Next, in Chap. 3, we present some technological challenges to designing and fabricating reliable nanoarchitectures, such as crossbar array design, variability-tolerant design, and ultra-low-power design. We then address LoC design and its specific application to diagnostics, food analysis, and environmental analysis. In particular, Chap. 4 describes technologies for on-chip sample transport and detection, while Chap. 5 presents gravimetric sensing methods. Chapter 6 describes in detail the application of LoCs to the detection of antibiotics in milk, as an example of the synergy of various technologies. Finally, Chap. 7 concludes the book by analyzing advanced technologies for nanoassembly and nanofabrication, including biological means to create nanostructures.

Advanced SoC Design

Advanced SoC design requires the use of new nanomaterials and devices, as well the appropriate choices of circuit architectures. In this section we give a brief introduction to nanodevices and to nanoarchitectural issues for advanced SoC design. Subsequent chapters will elaborate on some of these issues in more detail.

Devices: SiNW and CNT for Computational Electronics

Several early nanowire device technologies involve *bottom-up processing* techniques, i.e., device structures are formed from SiNW, by growing massive numbers of nanowires and later connecting them in functional topologies. In this case a large percentage of the grown structures may not be suitable for proper operation due to process parameter variations, structural irregularities, and alignment issues.

This approach requires mature self-assembly and self-test techniques to overcome the large device failure ratios and to exploit the functional redundancy in order to achieve the desired functionality. In addition to the difficulties of integrating reliable functions using bottom-up techniques, the approach also lacks a direct path to *hybrid integration* with mainstream silicon CMOS circuitry, which is an essential element for interfacing functional nanodevices with the external world.

One particular technology for nanowire device fabrication that has been attracting considerable interest in recent years involves *top-down* silicon-based processing of nanowires using modified lithographic techniques. The main advantage of the top-down technologies is that as they are defined by lithography, the wire placement on the wafer is accurately defined. Top-down SiNW technology fabrication is potentially no more challenging than the fabrication required for conventional bulk or SOI CMOS devices [5, 28, 29]. However, since most nanowires are sublithographic in nature, down-scaling and dimensional control issues are more challenging than for the bottom-up technologies. Another issue that might be problematic for top-down technologies is doping and the variability of the dopants in small dimensional structures, especially in cases where the processes defining the nanowires are influenced by the dopants, such as in the EPFL process. In Chap. 2 we explore some particularly promising approaches for the top-down lithographic fabrication of nanowire transistors and logic-on-a-wire structures that hold the promise of significantly improved performance in terms of current–voltage characteristics and direct integration with nanometer-scale CMOS devices.

An interesting technology for carbon nanotubes, developed at Stanford [32], uses silicon wafers with a layer of pseudo-aligned CNT that are transferred onto the surface with an experimental process step. These CNTs provide the means to form either *n* or *p* transistors, which benefit from both higher mobility and the lack of wells. Unfortunately, CNTs may be crooked and interrupted, and thus special processing steps need to be applied to yield reliable combinational logic cells that are reminiscent of their CMOS counterparts [6].

Devices: Ferroelectric Materials for Memories

Nonvolatile memory array applications constitute another major area of integrated nanosystems where nanoscale device technologies may offer specific functionality and integration possibilities that are otherwise not achievable with conventional technologies.

The quest for universal memory featuring high density, fast operation, and low power involves a large diversity of memory candidates such as MRAM (Magnetic RAM), PCRAM (Phase Change RAM), Fe-RAM (ferroelectric RAM), and molecular memories. Recently the ferroelectric property of some ultrathin films has been shown to be a good candidate for nonvolatile memory thanks to the bistability of polarization [35, 39]. In ferroelectric materials, the electrical dipoles orientate according to an applied external field and, after reaching a critical electric field, the resulting orientation can be maintained even when the field is switched off. This mechanism can be used to store charges and therefore to store information for nonvolatile memory applications.

As an example, we describe in Chap. 2 the processing and study of the properties of ultrathin P(VDF–TrFE) films when integrated in the gate stack of a MOSFET, which could operate as a one-transistor (1T) memory cell. This cell structure is more scalable than conventional 1T–1C memory architectures using ferroelectric capacitors. In addition to the unique device functionality, the explored technologies also lend themselves to hybrid integration with mainstream CMOS devices – and therefore, very valuable for large-scale nanosystem realizations.

Devices: SiNW, CNT, and Piezoelectric Materials for Integrated Sensing

While exploring top-down techniques for nanowire fabrication and integration with CMOS devices, we should also devote similar attention to various bottom-up technologies that can produce structures offering unique features for specific applications. In particular, sensor arrays based on grown nanowire (or nanotube) structures hold the promise of exploiting such unique functionality while also maintaining an acceptable level of immunity toward individual device failures through large-scale functional redundancy.

Recent examples of piezoelectric materials in polymer matrix have particularly versatile application areas, from ultrasound sensors in medicine, to underwater ultrasound in military and civilian applications, to smart sensor systems in automobiles, or to nondestructive testing in industry. Since the 1950s, lead titanate compounds ($PbTiO_3$) have been the leading materials for piezoelectric applications due to their excellent dielectric and piezoelectric properties. The rugged solid-state construction of industrial piezoelectric ceramic sensors enables them to operate under most harsh environmental conditions. However, the production of piezoelectric nanowires and their applications are still very challenging.

With the increasing public awareness of health concerns associated with lead and the recent changes in environmental policies, we have focused our research on the bottom-up growth/fabrication and characterization of $KNbO_3$-based nanowires. The nanowire arrays are subsequently embedded in a polymer matrix by immersion in an epoxy resin and the composite is easily peeled off of the $SrTiO_3$ substrate. The measured piezoactivity on such piezoelectric nanowire arrays is very encouraging for a very wide range of potential applications.

CNTs can be used to nanostructure electrodes used for protein sensing as they are amenable to efficiently capture the small amount of charge released, for example, by protein capture by an antibody. A study and use of CNTs for protein sensing is described in [7].

Circuit and Architectures: Fault Tolerance

Research on fault-tolerant circuits and architectures has been ongoing for decades, because it is motivated by the necessity of building highly dependable systems, independently of the underlying technology. Banks, insurance companies, and Internet service providers require *five nines* availability, i.e., 99.999% probability of being able to give a service at any given time despite any possible internal failure.

Some fault-tolerant computing technologies were developed for wafer-scale integration in the 1980s, under the assumption that the large number of devices on a wafer would support several types of redundancy. Nanotechnologies show some characteristics similar to wafer-scale integration. Indeed nanoscale circuit integration yields circuits with a large number of devices, and some of these devices can be used to boost reliability by means of appropriate redundancy schemes. Nevertheless, classic techniques such as triple-modular redundancy have high costs in terms of power consumption and should be used sparingly.

There are many approaches to achieving dependable nanosystem design; some rely on digital and others on analog circuit techniques. In the first class we can see the use of error detecting/correcting codes as well as the use of redundant devices and circuit blocks. In the second class, analog circuits can be used to compute digital functions by averaging and thresholding the results of analog computation.

Overall there is a broad tendency to use regular layout fabrics and a structured interconnect for nanosystem design. The most prominent example of regular fabric is the *crossbar architecture*, where regular rectangular grids support computation as well as switching and storage.

Crossbar Architectures

A crossbar is a rectangular mesh of wires designed in two orthogonal directions (Fig. 1.5). The crossbar is personalized by switching devices positioned at the cross-points; their pattern is often abstracted as the *personality matrix* of the array. The persistence and completeness of the personality distinguishes an array dedicated to computing from a read–write memory array. A computing array is dedicated to implementing a specific logic function; its personality is fixed and its dimensions are related to the function being implemented. Reconfigurable arrays add the twist of being able to change the personality during operation. Crossbar arrays are reminiscent of PLAs, often used in the 1970s because their regularity facilitated physical design (at a time when physical design was much less developed compared to now). PLAs lost ground to standard cells and other styles because of flexibility (inferior to standard cells) and the need for dynamic (or power-hungry pseudo-NMOS) operation in CMOS technology. Interestingly enough, Mo and Brayton recently revisited the use of PLAs with the objective of exploiting their regularity to achieve predictable timing [27]. De Hon [11] proposed the use of PLAs as computational structures for molecular electronics and nanowires. His work, deeply inspired by self-assembly, is based on the conjecture that you can create computational

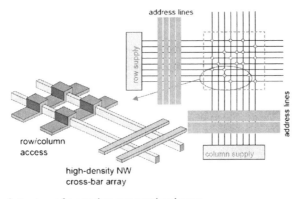

Fig. 1.5 General structure of a crossbar computational array

structures with a function unknown a priori and use later programming means to match the functional requirements.

When looking at future opportunities in nanoelectronics, it would be desirable to be able to use crossbars with connections (e.g., nanowires) that can have dimensions beyond the optical lithography limit. Nevertheless, a reasonable assumption is that the overall wiring structure in a SoC is still limited by lithography. Thus, an embedded nanoarray would have the appeal of being compatible with standard MTs, while having a smaller size and therefore higher computational density. Based on this conjecture, there are two important problems to be solved: (1) dealing with manufacturing defects and failure rates in the nanoarray and (2) interfacing the nanoarray to the external circuit. The former problem can be addressed by storing redundant information in various ways, as described in Chap. 3.

The latter interfacing problem is a new and critical problem presented by nanotechnologies. It encompasses the issues of connecting wires (called *meso-* and *nano*wires) of mismatching sizes and electrical driving strengths. Moreover, the external decoding and interconnection has to be such that no area is wasted around the nanoarray. In other words, if the interconnection structure is conservatively designed, then the area advantages of using a nanoarray disappears. Likharev and Strukov [23] proposed various interconnection schemes for nanoarrays based on a rotation of the nanoarray axes against the mesowire frame of reference to provide an efficient interconnection. A circuit architecture based on linear address decoders and multivalued logic is described in Chap. 3.

One of the major advantages of using crossbar architectures is that they achieve timing predictability due to the fact that wiring lengths can be easily computed within a rectangular array. In turn, timing predictability helps achieve fast design closure. Fast design closure is a prerogative of the synthesis tool flow that should allow a designer to complete a design with a limited number of changes to meet timing requirements. Clearly timing closure is closely related to timing predictability.

Networks on Chips

Timing predictability and design closure are also extremely important in the case of *on-chip* communication. The advancement of MT in terms of integration leads us to SoCs with many (e.g., 10 to 100) digital units (e.g., processor cores, controllers, storage, application-specific units) that need to be interconnected in an efficient and reliable way. Moreover, SoC architectures are often heterogeneous, i.e., units have different sizes and the communication requirements differ radically from point to point. The network on chip (NoC) technology developed rapidly in the first years of the new millennium [4, 13, 18] addresses three major design requirements: (1) realizing a modular and structured interconnect scheme, thus addressing predictability and timing closure issues; (2) overcoming the limitations of standard busses, which do not scale up in terms of connected components as far as performance and power consumption are concerned; (3) addressing reliability issues in the interconnect by providing path diversity as well as a layered approach to error detection and correction.

On-chip networks can be instrumental to connect crossbars and to realize a regular and programmable system architecture. Indeed we can envision a hierarchical NoC where the top layer is designed to match the application and system hardware resources. The lower layer provides the interconnect among the crossbars to realize functional and/or storage hardware resources, and it is directly tuned to the physical characteristics of the crossbars.

Analog Averaging

The severe parameter variations and structural irregularities observed in all nanoscale device fabrication technologies represent a significant challenge for the overall reliability and fault-tolerance of large-scale nanosystems. Not only are the failure rates much higher than what is seen in conventional integration technologies, but the increasingly higher stochastic nature of device behavior makes it more difficult to construct robust functional blocks out of inherently unreliable elements.

It has been shown that fine-grained redundancy techniques coupled with analog averaging of individual device responses offers significantly better results compared to conventional majority-voting techniques [36]. As an example, a digital function can be triplicated or quadruplicated, and the corresponding outputs be handled as analog signals that can be averaged and thresholded. Thus the final digital output enjoys the analog averaging of the intermediate signals. The range of potential problems extends not only to the correct identification of logic levels, but also to severe delay variability that can eventually render the entire system inoperable.

In this context, we explore a novel approach based on fine-grained redundancy and output averaging to optimize digital integrated circuit yield with regard to speed and area/power for aggressive scaling technologies. The technique is intended to reduce the effects of intradie variations using redundancy applied only on critical parts of the circuit. The work shows that the technique can be already applied to upcoming nanometer CMOS technology processes (65, 45, and 32 nm). However, the real benefit is expected for future nanoscale CMOS technologies where an optimal point is demonstrated in the speed vs. area/power tradeoff when using the new optimization technique. The explored design approaches are also very promising and immediately applicable for novel SiNW array realizations, where more complex functions are targeted. The technique is intended to reduce the effects of intradie variations using redundancy applied only on critical segments (critical paths). This way, the proposed technique can optimally be used in the design of large synchronous nanosystems.

Circuits and Architectures: Ultra-Low-Voltage Operation

Another very important aspect of complex nanosystem design using nanoscale fabrication technologies is the operation of such systems in very low (or ultralow) voltage ranges. The two main driving factors in this trend are (1) the physical constraints of the nanometer-scale devices that limit operation only to specific low voltage domains and (2) the need to save power in future, massively integrated nanosystems. However, low-voltage operation brings a host of challenges, most prominently in the area of variability: since device parameters exhibit a wide range of parameter variability, operation in low voltage ranges results in severe performance variability in terms of speed and leakage power dissipation. To overcome this problem, completely new and innovative circuit techniques will be needed. The design problem can in fact be generalized for a very wide range of different nanoscale device technologies, yet many of its key points can be demonstrated using nanometer-scale CMOS technologies as a test case.

In Chap. 3 we explore a new static circuit topology for subthreshold digital design. The proposed method, entitled adaptive VGS (AVGS) technique, provides strong control over the delay and active-mode leakage of the logic gates. It can be used for compensating variations and applied to any device or technology node. The proposed method is a good replacement for adaptive body biasing in emerging multigate devices that have a very small body factor.

Lab-on-a-Chip

A comprehensive description of LoCs goes beyond the scope of this book. (See [41] for details.) We will briefly describe here some issues and solutions in sample transport and capture, and then we describe applications in three domains: diagnostics, food analysis, and environmental analysis.

Sample transport can be achieved in different ways. Biological samples can be moved on chips by fluidic convection and by electric [14] or magnetic [21] means. Micro-/nanopumps can be realized on chip as well as channels on layers above functionalized silicon or amorphous material. Magnetic fields can move samples

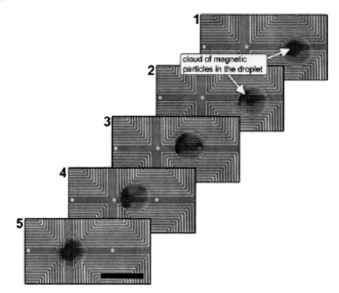

cloud of magnetic
particles in the droplet

Fig. 1.6 Droplet moved on a chip surface by magnetic fields [22]

that are attached to *microbeads*. As an example, Fig. 1.6 shows samples that are moved by means of a magnetic field generated by spirals designed on the top metal level of the chip [21, 22]. With this technology it is possible to achieve transport, split, and merge of droplets over a 2D array.

Sample analysis can be effected in specific areas of a LoC by bringing a sample in close proximity to a target. For example, a DNA strand can bind (i.e., hybridize) to a complementary probe. A conceptually similar – but more complex – mechanism can be used to trap proteins, e.g., by using antibodies as probes. With these working principles, *microarrays* can capture in parallel biological samples, thus becoming key instruments for *high-throughput biological* experiments [1, 15, 33] as well as medical diagnosis. The readout of microarrays can be done by optical or other means. Most current products use optical *labeled* techniques, i.e., samples are tagged with fluorophores, and the hybridized array is scanned optically, yielding a set of colored pixels to analyze. These techniques are bulky and hard to fully integrate in a monolithic chip. For this reason, there has been recent interest in *nonlabeled sensing*, i.e., techniques where the matching of a sample to a probe is measured by a variation of an electrical quantity (e.g., impedance, capacitance, current) that can then be measured by having the sensor under the probe itself. With this technology it is possible to integrate sensing with readout electronics and signal processing on the same substrate [19, 38]. Moreover, probes are organized as arrays, thus enabling parallel sampling.

Diagnostics

Many of today's high-sensitivity in vitro diagnostic testing procedures require both advanced sample fluidic transport and sensing. In the former domain, we consider in particular techniques that are based on handling magnetic micro- and nanoparticles (beads) due to their ability to separate molecules in low concentration from complex biological solutions or suspensions. Selection and resuspension of the beads is often based on the generation of a magnetic field and magnetic field gradients by mechanically moving or removing a permanent magnet in proximity to the biological liquid. These techniques can be integrated by using on-chip planar spirals to generate time-varying magnetic fields. Details are reported in Chap. 4.

Activities in this area also aim to demonstrate the use of diverse functions, such as magnetic actuation, microfluidics, optical, or acousto-gravimetric detection and their integration with signal processing. An important objective is to be able to detect on-chip small amounts of samples or weak concentrations of (bio)chemical compounds such as DNA, hormones, or proteins present in complex biological fluids (i.e., blood or plasma). The detection resolution is expected to be better than the performance of state-of-the-art optical detection methods like enzyme-linked immunosorbent assays (ELISA) in microtiter plates. By transporting the liquid sample through a suspension of magnetic nanoparticles retained within a microfluidic structure, it is possible to develop an assay with integrated optical detection on-chip that will provide a sensitivity of a few picograms per milliliter, at a fraction of the time of current ELISA tests. Overall, the application of sensing to diagnostics, possibly in conjunction with wireless connection, is extremely important to develop implanted and point-of-care analysis tools for medical practice [30].

In this area, it is also important to develop acousto-gravimetric transducers with a radical improvement of sensitivity and resolution for label-free biomedical sensors. The aim is to enhance diagnostic capabilities with regard to cost and disposability. Today, the most advanced acousto-gravimetric sensors achieve sensitivities of a few nanograms per milliliter of antigen concentration. This is not quite sufficient for widespread and universal use in immunoassays and other diagnostics. These issues are described in detail in Chap. 5.

Food Analysis

An important application of LoCs is the detection of contaminants in agricultural raw materials (e.g., cereals, grapes/wine, and other fruits), as well as in milk, meat, and processed foods. These LoCs will either replace cumbersome and costly laboratory methods or will be the basis of new screening methods. One of the priorities is having access to specific molecular probes for the targeted contaminants, such as DNA probes for toxigenic fungi, as well as antibodies and artificial receptors for mycotoxins, antibiotics, and pesticides. Parallel transduction technologies for detection (optical, electrical, acoustic wave, etc.) can be applied to detect several contaminants simultaneously. Microfluidics in the cartridge format and the use of magnetic actuation simplify sample preparation and reduce reaction and analysis times. Microsystem technologies enable the development of miniaturized, multifunctional, and portable systems that can even be envisaged for field applications.

A set of compounds of vital importance have been identified as targets for detection. Chemical substances (antibiotics, pesticides, and mycotoxins) and living organisms (toxigenic fungi, pathogens) can be considered as presenting the most relevant issues to be addressed from the point of view of safety. More specifically, the compounds and organisms to be detected include, but are not limited to, antibiotics, pesticides, mycotoxins, and pathogens. Detection can be achieved using receptors, antibodies, and DNA according to the circumstances.

As an example of LoC technology, Chap. 6 describes the detection of antibiotics in milk. This chapter shows the combination of various technologies, ranging from microfluidics to optical sensing and detection.

Environmental Analysis

Several techniques exist and are used in the field of environmental analysis to detect some pollutants. Nevertheless they usually involve sample collection, transportation, and laboratory analysis. We believe that improved sampling and subsequent processing in miniaturized fluidic systems leads to shorter analysis times, which will clearly improve the response time of the authorities and enable better prevention and better territory planning. In addition, pathogen detection could be an effective solution to monitoring or even preventing the spread of dangerous viruses in domestic and imported animal populations. Interesting detection targets for the environment are carbon-based microparticles, spores, viruses, and bacteria. This activity is not reported in detail in this book. Nevertheless it is an ongoing research topic of the nano-tera.ch initiative [30].

Advanced Fabrication Technologies for Nano-Bio Systems

The manufacture of electronic integrated circuits and systems is traditionally based on optical lithography. Technological improvements, such as using immersion lithography and optical proximity correction techniques, enable foundries to mass fabricate circuits reliably in the 45-nm technology node using 193-nm steppers. When going below 45 nm, an open question relates to the possible patterning means, which range from advanced optical (deep UV–X-ray) lithography to self-assembly methods.

When considering nanomaterials, such as silicon or gold nanowires, the objective to assemble deeply scaled lines (around and below 10 nm in pitch) or crossbars is extremely challenging. Chapter 7 explores different paths to the fabrication of nanowires. Note that these nanowires are used mainly as structures to support biosensing. Thus, the material of choice is often gold, even though some techniques can be extended to other materials.

Nanowire fabrication can be achieved via top-down (lithography-based) and bottom-up (aggregation-based) methods, as well as with a combination of both. In particular, nanowires can be prepared using extreme ultraviolet interference lithography (EUV-IL), developed at the Paul Scherrer Institute in Switzerland. Advanced lithography methods can be enhanced by other experimental nanofabrication steps, such as DNA-assisted assembly of gold nanoparticles and nanorods onto templates produced by EUV-IL. Alternatively, it is also possible to fabricate nanowires in various materials (e.g., Au, Al, Pt, Ag, ITO) through etching, lift-off, and shadow-deposition-based methods. Line widths down to 15 nm and periods down to 30 nm can be produced over areas of about 1 to 10 mm^2. Finally, nanowire fabrication can be finished by various means, including annealing for stabilization of the nanostructure and control of electronic properties.

Nanowire Characterization

The properties of nanowires determine their performance. Thus it is important to characterize nanowires in both the mechanical and electrical domains.

The nanoscale structure of metallic systems strongly influences their electronic and mechanical properties. Transmission electron microscopy is used for the characterization of the microstructure in order to understand the properties of metallic systems and to provide feedback to the production process. For gold nanowires, the mechanical properties are determined as a function of temperature and strain rate. The nanostructure of the material is the starting point for analysis.

When characterizing electronic properties, a key parameter is electrical conductivity as a function of downscaling the geometrical dimensions. An important objective is to determine the minimal dimension for gold wires to be used in nanoelectronics. The general mechanical properties of gold nanowires are also very important, such as the specific influence of mechanical deformation (elastic and plastic) on electrical conductivity. In addition, it is important to determine how and when electromigration is an issue for nanoelectronics as a failure mechanism.

Applications of Nanowires to Bioelectronics

Nanowires have several applications as nanoelectrodes and more generally in bio-nanotechnology. Nanowires are useful in this domain because of their relatively small diameter. Thus the electrical conductivity in nanowires is subject to quantum confinement and conduction is in separate quantized energy levels.

As a result, nanowires are attractive bioelectrochemical transducers since their diameters match the size of biochemical analytes and since their conduction is sensitive to surface perturbations. Thus nanowires have been incorporated into devices for biosensing purposes, such as PH measurement, DNA and protein detection, and in markers of various types.

Cooperative Engineering

Cooperative engineering is a key factor in realizing SoCs and LoCs, and especially those that monitor biological information or interface with living beings. Indeed the required technical skills to design and operate such systems are shared among engineers, computer scientists, chemists, physicists, biologists, and medical doctors. It is very important to find ways of translating specific technical idioms and to provide means for researchers with different backgrounds to communicate. For these reasons, the abstraction and modularity of information about system design and operation is extremely important. In the past, DT for integrated systems were encompassed in hardware/software codesign techniques. This notion has to be generalized to the concurrent design of complex multifaceted systems. The ability to determine models and interfaces among various system aspects is a key aspect of design and DT of the future.

At present, a few multidisciplinary research programs are tackling the design of heterogeneous systems (embedding SoCs and LoCs) and the related technologies. Among these, the Swiss *nano-tera.ch* [30] program addresses the improvement of human health and monitoring the environment by developing micro/nano/info technologies that enable one to design and manage distributed embedded systems. This program is intentionally multidisciplinary and multi-institutional, with the objective of gathering wide expertise in both technologies and application areas. This program grew out of the experience of the Swiss Competence Center for Materials CCMX/MMNS research effort, whose results are described in this book. It will also provide a vehicle for continuing this research path.

Summary

This chapter introduces a set of research topics on advanced SoCs, LoCs, and related technologies. These issues have been addressed by a group of researchers who worked as a team to address complex issues, such as the frontier of scaling for SoCs and the ultimate hybridization of technologies for LoCs.

This research work was motivated by the search for materials and devices that could benefit new systems, products, services, and markets. This research has shown that nanotechnology is having an ever-increasing impact on integrated system design and that its effective use can only be gauged by exploring holistically the path from materials to systems/services that include intermediate steps as shown in Fig. 1.1. Indeed the advantages (and disadvantages) of the use of particular materials can by exploited (and offset) by the judicious choice of circuits and architectures to achieve the overall system objectives in terms of performance, precision, cost, power consumption, and dependability.

References

1. http://www.affymetrix.com/index.affx
2. Allen J (1995) Natural language understanding. Benjamin Cummings, San Antonio
3. Bake D, Church G, Collins J, Endy D, Jacobson J, Keasling J, Modrich P, Smolke C, Weiss R (2006) Engineering life: building a fab for biology. Sci Am 294(6):44–51
4. Benini L, De Micheli G (2002) Networks on chip: a new design paradigm. IEEE Comput 35:70–78
5. Ben Jamaa MH, Moselund KE, Atienza D, Bouvet D, Ionescu MA, Leblebici Y, De Micheli G (2008) Variability-aware design of multilevel logic decoders for nanoscale crossbar memories. IEEE Trans Comput Aided Des Integr Circuits Syst 27(11):2053–2067
6. Bobba S, Zhang J, Pullini A, Atienza D, Mitra S, De Micheli G (2009) Design of compact imperfection-immune CNFET layouts for standard-cell-based logic synthesis. Design, Automation and Test in Europe, DATE 09, 2009, pp. 616–621
7. Carrara S, Shumyantseva VV, Archakov AI, Samorì B (2008) Screen-printed electrodes based on carbon nanotubes and cytochrome p450scc for highly-sensitive cholesterol biosensors. Biosens Bioelectron 24:148–150
8. Cerofolini G (2007) Realistic limits to computation II: the technological side. Appl Phys A Mater Sci Process 86(1):31–42
9. Close GF, Wong H-SP (2007) Fabrication and characterization of carbon nanotube interconnects. IEEE International Electron Devices Meeting (IEDM), Washington, DC, 10–12 December 2007, pp 203–206
10. Cox RV, Kamm CA, Rabiner LR, Schroeter J, Wilpon JG (2000) Speech and language processing for next-millennium communications services. Proc IEEE 88(8):1314–1337
11. de Hon A (2003) Array-based architecture for FET-based nanoscale electronics. IEEE Trans Nanotechnol 2(1):23–32
12. De Micheli G (1994) Synthesis and Optimization of Digital Circuits. McGraw-Hill, Columbus
13. De Micheli G, Benini L (2006) Networks on Chip. Morgan Kaufmann, San Francisco
14. Demierre N (2008) Continuous-flow separation of cells in a lab-on-a-chip using liquid electrodes and multiple-frequency dielectrophoresis. PhD Thesis, Lausanne
15. De Risi J, Penland L, Brown P, Bittner M, Meltler P, Ray M, Chen Y, Su Y, Trent M (1996) Use of a cDNA microarray to analyze gene expression patterns in human cancer. Nat Genet 14(4):457–460
16. Ecoffey S, Mazza M, Pott V, Bouvet D, Schmid A, Leblebici Y, Declercq MJ, Ionescu AM (2005) A new logic family based on hybrid MOSFET-Polysilicon nano-wires. IEEE International Electron Device Meeting, Washington, DC, December 2005
17. Ecoffey S, Pott V, Bouvet D, Mazza M, Mahapatra S, Schmid A, Leblebici Y, Declercq MJ, Ionescu AM (2005) Nano-wires for room temperature operated hybrid CMOS-NANO integrated circuits. Digest of Technical Papers IEEE International Solid-State Circuits Conference, 6–10 February 2005, pp 260–262
18. Guerrier P, Greiner A (2000) A generic architecture for on-chip packet-switched interconnections. Design Automation and Test in Europe Conference, Paris, France, March 2000, pp 250–256
19. Guiducci C, Stagni C, Zuccheri G, Bogliolo A, Benini L, Samorì A, Riccò B (2004) DNA detection by integratable electronics. Biosens Bioelectron 19:781–787
20. http://www.itrs.net/
21. Lehmann U, Sergio M, Pietrocola S, Niclass C, Charbon E, Gijs MAM (2007) 14th International Conference on Solid-State Sensors, Actuators and Microsystems, Transducers'07 and Eurosensors XXI, Lyon, France, 10–14 June 2007

22. Lehmann U (2008) Manipulation of magnetic microparticles in liquid phases for on-chip biomedical analysis methods. Ph.D Thesis, EPFL
23. Likharev KK, Strukov DB (2004) Introducing molecular electronics. Springer, Berlin
24. Manz A, Graber N, Widmer HM (1990) Miniaturized total chemical analysis systems: a novel concept for chemical sensing. Sens Actuators B 31:244–248
25. Maslov D, Falconer SM (2008) m. Mosca, 'quantum circuit placement'. IEEE Trans CAD 27(4):752–763
26. Mihic C, Simunic T, De Micheli G (2007) Power and reliability management of socs. IEEE Trans VLSI 15(4):391–403
27. Mo F, Brayton R (2002) Whirlpool plas: a regular logic structure and their synthesis. Proc ICCAD, pp 543–550
28. Moselund KE, Pott V, Bouvet D, Ionescu AM (2008) Hysteretic inverter-on-a-body-tied-wire based on less-than-10mv/decade abrupt punch-through impact ionization MOS PIMOS switch. Proceedings of International Symposium on VLSI Technology, Systems and Applications (2008 VLSI-TSA), Taiwan, 21–23 April 2008
29. Moselund KE, Bouvet D, Ben Jamaa MH, Atienza D, Leblebici Y, De Micheli G, Ionescu MA (2008) Prospects for logic-on-a-wire. Microelectronic Eng 85(5–6):1406–1409
30. http://www.nano-tera.ch
31. Paradiso J, Starner T (2005) Energy scavenging for mobile and wireless electronics. IEEE Pervasive Comput 4(1):18–27
32. Patil N, Jie D, Wong H-SP, Mitra S (2007) Automated design of misaligned-carbon-nanotube-immune circuits. Proceedings of the Design Automation Conference, June 2007, pp 958–961
33. Pease A et al (1994) Light-generated oligonucleotide arrays for rapid dna sequencing analysis. Proc Natl Acad Sci 91(11):5022–5026
34. Pedram M, Rabaey J (2002) Power aware design methodologies. Springer, Berlin
35. Salvatore GA, Bouvet D, Stolitchnov I, Setter N, Ionescu AM (2008) Low voltage ferroelectric FET with sub-100nm copolymer P(VDF-trfe) gate dielectric for non-volatile 1T memory. ESSDERC 2008, Edinburgh, Scotland, 15–19 September 2008
36. Schmid A, Leblebici Y (2004) Robust circuit and system design methodologies for nanometer-scale devices and single-electron transistors. IEEE Trans VLSI 12(11):1156–1166
37. Shende VV, Prasad AK, Markov IL, Hayes JP (2003) Synthesis of reversible logic circuits. IEEE Trans CAD 22(6):710–722
38. Stagni C, Esposti D, Guiducci C, Paulus C, Schienle M, Maugustyniak, Zuccheri G, Samori B, Benini L, Ricco B, Thewes R (2006) Fully electronic CMOS DNA detection array based on capacitance meausurement with on-chip analog to digital conversion. Proceedings ISSC, San Francisco, 2006, pp. 69–78
39. Stolichnov I, Riester SWE, Trodahl HJ, Setter N, Rushforth AW, Edmonds KW, Campion RP, Foxon CT, Gallagher BL, Jungwirth T (2008) Non-volatile ferroelectric control of ferromagnetism in (Ga, Mn)As. Nat Mater 7(6):464–467
40. http://www.tilera.com
41. Tüdos A, Besselink G, Schasfoor X (2001) Trends in miniaturized total analysis systems for point-of-care testing in clinical chemistry. Lab Chip 1:83–95
42. Vangal SR et al (2007) An 80-Tile 1.28TFLOPS network-on-chip in 65nm CMOS. Proceedings of the International Solid-State Circuits Conference, 11–15 Febrauary 2007, pp 98–99
43. Vangal SR, Howard J, Ruhl G, Dighe S, Wilson H, Tschanz J, Finan D, Singh A, Jacob T, Jain S, Erraguntla V, Roberts C, Hoskote Y, Borkar N, Borkar S (2008) An 80-tile sub-100-W teraflops processor in 65-nm CMOS. IEEE J Solid State Circuits 43(1):29–41
44. Vankamamidi V, Ottavi M, Lombardi F (2008) Two-dimensional schemes for clocking/timing of QCA circuits. IEEE Trans CAD 27(1):34–44

Chapter 2
Materials and Devices for Nanoelectronic Systems Beyond Ultimately Scaled CMOS

Didier Bouvet, László Forró, Adrian M. Ionescu, Yusuf Leblebici,
Arnaud Magrez, Kirsten E. Moselund, Giovanni A. Salvatore, Nava Setter,
and Igor Stolitchnov

Introduction

The technological and economic feasibility of high-density, large-scale nanoelectronic systems integration is still being driven by the fundamental paradigms of classical CMOS technology, for which there will be no apparent substitutes in the next 10 to 15 years. However, we cannot expect to continue the lithography scaling of classical CMOS devices and circuits indefinitely, due to fundamental physical limitations such as process variability, excessive leakage, process costs, and very high power densities. This observation calls for radical action on several fronts in order to ensure the continuity of the nanoelectronic systems integration paradigm *until* one or more feasible alternative technologies emerge to replace CMOS within the next 15-year time frame. In particular, we will have to consider the introduction of *new materials and technology steps* to augment and ameliorate the classical CMOS process/devices, explore *new device structures* that can provide reliable performance and sufficiently low power dissipation at high-density integration, and develop new fabrication technology for the CMOS-compatible integration of new materials and nanostructures for new devices in CMOS process flows (approach of stepwise substitution).

In order to keep a strong link to existing CMOS technologies and the ITRS roadmap, compatibility with silicon-based fabrication technologies shall be an important guideline, at least in the near term. On the one hand, this calls for *new materials and fabrication techniques* to further improve the device characteristics and to overcome the limitations of the existing and projected CMOS devices (such as high-k dielectrics, high-mobility substrates, etc.). On the other hand, new nanotechnology components will have to be *hybridized with silicon CMOS* as an add-on approach. Ultimately, the road to developing completely revolutionary nanotechnology platforms (including materials, devices, and integration technologies) should be open – yet the near-term approach will also require a closer link with silicon-based CMOS technologies.

Yusuf Leblebici (✉)
EPFL – Swiss Federal Institute of Technology, Lausanne, Switzerland

G. De Micheli et al., *Nanosystems Design and Technology*,
DOI 10.1007/978-1-4419-0255-9_2, © Springer Science+Business Media LLC 2009

Clearly, the stated objectives can only be achieved with a strong collaboration and interplay of all disciplines involved in this endeavor to search for new solutions. In particular, the innovations to be explored at the circuit and system levels will have to be inspired by the possibilities offered by new materials and device technologies – while the direction of materials and device research will have to be influenced by the architectural choices made at the system level.

In this chapter, we review some of the most recent results in these areas and put them in a unified context that covers a very wide range, from materials to system design. The first section presents a top-down silicon nanowire fabrication platform for high-mobility gate-all-around (GAA) MOSFETs and impact-ionization devices. Ferroelectric FET with sub-100-nm copolymer P(VDF-TrFE) gate dielectric are examined in the next section for nonvolatile memory applications, which is a very promising direction toward future high-density memory arrays, followed by a discussion of materials for piezoelectric nanodevices in the last section.

Top-Down Silicon Nanowire Fabrication Platform: High-Mobility Gate-All-Around MOSFETs and Impact-Ionization Devices

The main advantage of the top-down technologies are that as they are defined by lithography, the wire placement on the wafer is accurately defined. Top-down silicon nanowire technology fabrication is potentially no more challenging than fabrication of conventional bulk or silicon-on-insulator (SOI) CMOS devices [1, 2]. However, since most nanowires are sublithographic in nature, down-scaling and dimensional control are more challenging than for the bottom-up technologies. Another issue that might be problematic for top-down technologies is doping and the variability of the dopants in small-dimensional structures, especially in cases where the processes defining the nanowires are influenced by the dopants, as is the case in our process. Oxidation rates are highly dependent on the doping concentration and, to a lesser degree, on the etching. However, doping variability is an issue for all nanoscale electronics.

In this work we have developed a versatile top-down silicon nanowire platform that allows for the fabrication of various devices, reported in detail in [3–11]. We have demonstrated enhanced electron mobility due to oxidation-induced tensile strain in GAA bent MOSFETs. We have developed punch-through impact-ionization MOS (PIMOS) devices on body-tied Ω-gate MOS structures. These devices show less than $-10\,\mathrm{mV/dec}$ abrupt switching combined with abrupt switching in both $I_D(V_{DS})$ and $I_D(V_{GS})$. A 1-PIMOS DRAM memory cell has been demonstrated, whereas we have also fabricated an NMOS inverter that maintains an abrupt transition and hysteresis and shows a unique gain of -80.

By combining the resources of technology and device research with that of circuit design, we have demonstrated the ability to create some innovative electronic building blocks that represent the first steps toward logic-on-wire. In the future we would like to extend this work even further and experimentally demonstrate more advanced nanowire architecture such as the addressing scheme of crossbar memories.

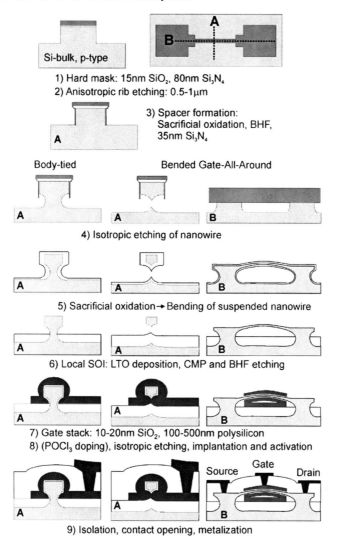

Fig. 2.1 Process flow for top-down fabrication of silicon nanowires with pentagonal cross-section. Cross-section A is shown for both body-tied and suspended (bended GAA) devices, whereas the cross-section B is only shown for the suspended wires

Top-Down Nanowire Fabrication Platform

A schematic of the process flow is shown in Fig. 2.1; it depicts devices of pentagonal cross-section, but a triangular cross-section is obtained by varying the anisotropic rib etching, step 2, and the isotropic etching, step 4. Two different cross-sections are shown for the electronic device. The first one, A, is across the MOSFET channel, and the second, B, runs along the MOSFET channel from the source plot to the drain plot. For the body-tied device, only cross-section A is shown, whereas for the

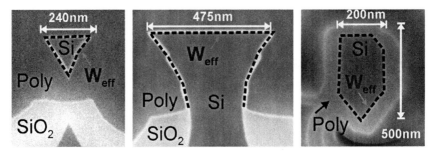

Fig. 2.2 Cross-section SEM/FIB images of triangular and tri-gate fabricated devices. Also illustrated: the definition of effective width, W_{eff}, of three different devices

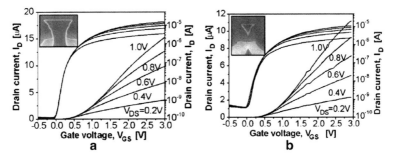

Fig. 2.3 $I_D(V_{GS})$ characteristics for 10-μm-long devices; *inset* shows a corresponding cross-section. **a** Tri-gate device with $W_{\text{eff}} = 1.12\,\mu$m. **b** Gate-all-around device with $W_{\text{eff}} = 400\,$nm

suspended bended GAA device both cross-sections A and B are of interest. The triangular devices are primarily developed for field-effect transistors and fabricated on p-type (boron) (100) wafers of resistivity 0.1 to 0.5 cm, which corresponds to a doping density of $3 \times 10^{16}\,\Omega\,\text{cm}^{-3}$ to $2 \times 10^{17}\,\text{cm}^{-3}$. Devices with a pentagonal cross-section, as shown in the process flow in Fig. 2.1, are fabricated on high-resistivity p-type (boron) wafer 15 to 25 Ω cm, $5 \times 10^{14}\,\text{cm}^{-3}$ to $9 \times 10^{14}\,\text{cm}^{-3}$, and they serve punch-through impact-ionization (PIMOS) devices. These latter devices are also made with triangular cross-sections. The triangular process flow is simpler; in this case surface roughness on the vertical side walls is not an issue, and the aspect ratio is smaller, which makes the local-SOI fabrication easier. Figure 2.2 shows cross-sections of some of the fabricated devices using the process from Fig. 2.1.

Devices of a wide range of cross-sections have been fabricated and characterized. Device effective widths range from 16 nm, for the smallest circular GAA MOSFET, to 40 μm, for quasi-planar devices, and gate lengths of characterized devices range from 0.9 to 20 μm. Conventional $I_D(V_G)$ characteristics of typical $10 - \mu$m-long fabricated devices with two different sections are shown in Figs. 2.3 and 2.4. The I_{on} current measured at $V_{GS} = V_{DS} = 2\,$V is approximately 22 μA/μm for the GAA MOSFET and about 10 μA/μm for the tri-gate device, in both cases for a 10-μm-long device. The corresponding values for I_{off} measured at $V_{GS} = 0\,$V

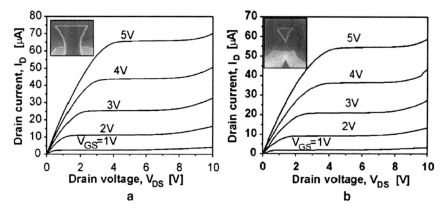

Fig. 2.4 $I_D(V_{DS})$ characteristics for the devices reported in Fig. 2.3

and $V_{DS} = 0.2$ V is around 0.15 nA/μm for both devices. The reason behind the enhanced current drive of the GAA devices is related to process-induced local tensile strain (bended channel) and mobility enhancement. The flat $I_D(V_{DS})$ characteristics in Fig. 2.4 are also evidence that they do not suffer from self-heating effects because drain and source are connected to the silicon bulk. The absence of any kink in the output characteristics indicates that all devices have a fully depleted body.

Bended Gate-All-Around MOSFET

The bended GAA MOSFET is a new device concept proposed in [7], where the transistor channel is bended by thermal oxidation, introducing the benefits of improved carrier mobility due to the local tensile strain. Figure 2.5 illustrates typical bending as observed in our top-down fabricated nanowire. It is worth noting that bending remains even after fabrication of the gate stack (here 10 nm SiO_2 and 100 nm polysilicon) around the silicon wire.

Micro-Raman spectroscopy has been used for evaluating and confirming tensile strain in our suspended nanowire structures based on a setup adapted for nanoscale measurements developed at the University of Newcastle. Typical measurements showed strain values around 1.5%, which corresponds to approx. 2 GPa, whereas the greatest measured strain is 3.2% (∼4 GPa).

The $I_D(V_{GS})$ characteristics, normalized per effective width, for two bended GAA MOSFETs, compared to that of a nonbended tri-gate device, are shown in Fig. 2.6. The relative current enhancement for device A is a function of gate overdrive varying between 80 and 160% (highest value close to the threshold voltage). Figure 2.7 shows the extracted low field mobility as a function of effective width, W_{eff}, for devices of three different channel lengths. For classical devices there should not be any dependence of the mobility on effective width, which is also what we

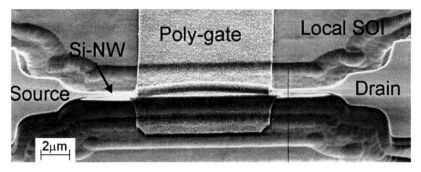

Fig. 2.5 Illustration of bended gate-all-around MOSFET (the poly gate is short-circuited on the two lateral sides to underlying poly line)

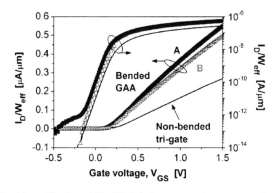

Fig. 2.6 Normalized (per effective width, W_{eff}) drain current, I_D, vs. gate voltage, V_{GS}, in bended and nonbended (body-tied tri-gate) nanowire MOSFETs with $L = 5\,\mu\text{m}$, at $V_{DS} = 20\,\text{mV}$. Effective widths are A: $0.56\,\mu\text{m}$, B: $0.62\,\mu\text{m}$ and tri-gate: $1.32\,\mu\text{m}$

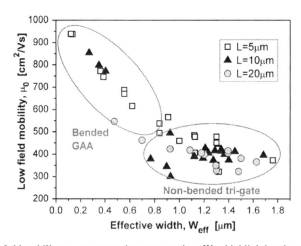

Fig. 2.7 Low-field mobility, μ_0, vs. nanowire cross-section, W_{eff}, highlighting the effect of tensile-strain-induced mobility enhancement in bended structures (for $W_{\text{eff}} < 2\,\mu\text{m}$), at room temperature

can see for the larger nonbended devices. For the smaller bended devices one can observe a rapid increase in the mobility with decreasing dimension. For the smallest devices we see an increase in mobility of around 100%. These findings are well correlated with the Raman strain measurements, where an increase in the tensile strain was observed with decreasing dimension, and tensile strain has been known to cause improvements in the electron mobility of similar magnitude for this amount of strain. This demonstrates that high-mobility channels can be obtained in bended GAA MOSFET.

Impact-Ionization Devices and Inverter on Silicon Rib

The investigated PIMOS device structure is represented in Fig. 2.8a; it is designed as a conventional NMOS, in contrast to a previous I-MOS device [12] that is a reversed biased gated pin structure. PIMOS has a very low-doped body/channel region ($N_A \sim 6 \times 10^{15}$ cm^{-3}) and fairly abrupt $N+$ region (arsenic doped, 10^{20} cm^{-3}). The body of the device is shaped as an X- or tri-gate for small widths, W, and develops gradually into a quasi-planar device for large W. For low drain bias the PIMOS operates as a conventional MOSFET (see $V_{DS} = 0.2$ V in Fig. 2.5b), as V_{DS}'s increased impact ionization will take place at the drain-substrate junction and the drain current will increase rapidly (see $V_{DS} > 8.3$ V in Fig. 2.5b); so far this could resemble the operation of a MOSFET at very high drain bias. However, if at the moment when avalanching sets in the condition of punch-through is met and the device is biased in subthreshold, the particular PIMOS functionality will set in because the electrostatic potential in the channel creates a saddle point in the channel near the drain junction. The electrons generated by impact ionization are swept into the drain junction, whereas the holes can be separated into two fractions. Part of the holes will migrate toward the point of lowest potential, creating a hole pocket at the interface; this also causes a displacement away from the interface of the electron current (Fig. 2.8a). The filling of the hole pocket at the saddle point will result in an increase in the body potential, and thereby a decrease in the local threshold voltage as a function of the drain potential. Another part of the holes constitutes a substrate current, which causes a voltage drop across the substrate resistance that will eventually forward bias the source-substrate junction and turn on the parasitic bipolar structure. The bipolar gain will result in a further amplification of the base (substrate) current, and hence also the channel current. This forms a positive feedback loop on the current, with an abrupt increase in the drain current and thus a steep inverse subthreshold slope.

The hysteresis observed in both the $I_D(V_{GS})$ (Fig. 2.8b) and the $I_D(V_{DS})$ characteristics is due to a combination of the impact ionization under the high drain bias and the positive feedback loop sustained by the BJT action. When the positive feedback loop can no longer be maintained, the drain current will drop suddenly.

Table 2.1 summarizes the main figures of merit of PIMOS compared to the I-MOS (proposed by Stanford University). For this comparison a lateral I-MOS is

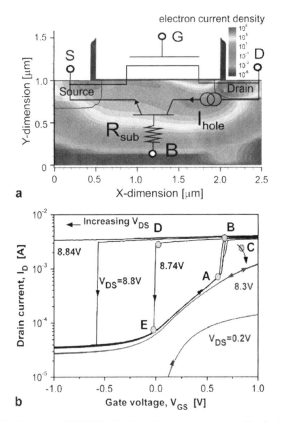

Fig. 2.8 **a** TCAD simulation (DESSIS) showing the electron current distribution at breakdown in a PIMOS device with the schematic of the parasitic bipolar in parallel. The hole pocket at the interface pushes the electron current toward the bulk. **b** Experimental abrupt switching and hysteresis in $I_D(V_{GS})$ for a PIMOS device of $L = 1.4\,\mu m$ and $W = 10\,\mu m$. Increasing V_{DS} widens the hysteresis by displacing the *down* transition $D \rightarrow E$, whereas the *up* transition $A \rightarrow B$ remains more or less fixed. Two curves for normal MOS operation are shown with $V_{DS} = 0.2$ and 8.3 V, respectively

considered, as it is the most closely related in terms of technology and integration with CMOS. The PIMOS fabrication is more closely related to CMOS, and good reliability is also obtained at higher temperatures. The challenges for the PIMOS device are in terms of control of the subthreshold leakage current and the inherent short channel effects. The hysteresis effect can be an advantage for certain applications (memory), but its control, or even elimination, is required for logic applications. The I-MOS, on the other hand, presents a much lower leakage than a corresponding MOSFET, and the scalability in terms of bias voltages is therefore limited by material parameters. The main challenges for the I-MOS are in terms of processing technology and reliability. In planar devices, operation is only assured for a few cycles, but this can be solved by the use of a vertical structure.

Table 2.1 Comparison of main figures of merit of PIMOS and I-MOS

	PIMOS	I-MOS
I_{off}	$I_{leak} = I_{D0}e^{-(V_T/S)}$, high	$I_{leak} = I_s = \left[\frac{eD_p p}{L_p} + \frac{eD_n n}{L_n}\right]$, low
SCE	Inherent	Better than CMOS *off* mode
Slope $I_{off} \rightarrow I_{on}$	<10 mV/dec	<10 mV/dec
Critical dimension	L_{eff}	$L_{gate} + L_I$
Processing	Self-aligned	More difficult
Hysteresis	Yes	No
Cycling	>10^4 cycles	Few cycles
Substrate	Bulk	SOI
Immunity to I_{sub}	Bad, requires wells	Good, fabricated on SOI
Scalability of V_{BD}	Limited by I_{off}	Limited by material
Ultimate scalability	Limited by I_{off}	Limited by lithography
Temperature stability	Excellent	Unknown

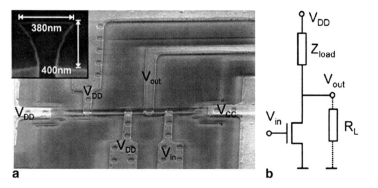

Fig. 2.9 a SEM image of NMOS inverter with two-transistor resistive load on a body-tied wire.
b Equivalent schematic of biasing scheme

We have demonstrated for the first time an N-PIMOS inverter consisting of a PIMOS input device and a pull-up NMOS load (Fig. 2.9), all fabricated on a single etched silicon rib (or body-tied silicon wire with sub-micron cross-section).

The inverter reproduces the abrupt switch and hysteresis from the $I_D(V_{GS})$ characteristics; the voltage transfer characteristic (VTC) is shown in Fig. 2.10. The PIMOS inverter VTC is different from both that of a conventional CMOS and I-MOS inverters. VTC of PIMOS incorporates three distinct regions of operation: (a) when the input PIMOS device is off, the pull-up network will pass the value of VDD to the output, (b) when $V_{DS} > V_{DBD}$ when the input exceeds the gate pull-up voltage, $V_{in} > V_{GPU}$, the PIMOS will pull down the output abruptly until it reaches the drain breakdown voltage, and (c) beyond V_{DBD} the device is still on, unlike an I-MOS, it just works as a quasi-conventional short-channel MOSFET at high V_{DS}, after abrupt switching in V_{GS} where the current I_D is dictated by both MOSFET and bipolar contributions. So the MOSFET operation will continue to pull down the

Fig. 2.10 Voltage transfer
characteristics of PIMOS
inverter reported in Fig. 2.5

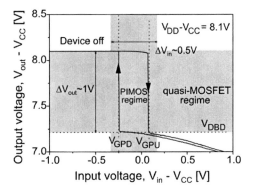

Fig. 2.10 Voltage transfer characteristics of PIMOS inverter reported in Fig. 2.5

output until the point determined by the load network, i.e., a resistive divider in the case of an N-PIMOS.

The hysteretic VTC characteristic in Fig. 2.10 also supports the basic idea of an SRAM cell made of a single inverter (and possible to fabricate on a single etched silicon wire), especially because the hysteresis width can be optimized by the value of applied V_{DS}. However, the main drawback of such a memory cell is the high current consumption. Future prospects should concern the demonstration of a complementary PIMOS inverter to reduce the static power consumption.

Discussion

The work was divided into two distinct topics each dealing with a different issue: (a) mobility enhancement in a GAA architecture and (b) small slope switching based on impact ionization. However, they are united by a common technology platform, which allows for the low-cost fabrication of 3D devices with sublithographic dimensions. Another common feature is that all these devices can be described as true 3D structures, because of their nonplanar geometry and 3D electrostatic effects.

We have addressed the issue of strain-induced mobility enhancement in silicon nanowires. By bending the wire as a result of one or more sacrificial oxidation steps we have demonstrated a bended MOSFET with tensile strain on the order of 1 to 3% and a mobility improvement of up to 100%.

We have proposed a punch-through impact-ionization device (PIMOS) based on a low-doped MOSFET structure operated in subthreshold under punch-through conditions. With this device we obtain abrupt switching transients of less than 10 mV/dec due to the impact-ionization mechanism. In addition, as a result of the inherent parasitic bipolar action, we observe a bias-dependent hysteresis in both the $I_D(V_{DS})$ and $I_D(V_{GS})$ characteristics. We have demonstrated an abrupt hysteretic inverter based on the PIMOS $I_D(V_{GS})$ hysteresis. The N-PIMOS inverter has three operating regions; off, PIMOS mode, and normal MOSFET mode.

Ferroelectric FET with Sub-100-nm Copolymer P(VDF-TrFE) Gate Dielectric for Nonvolatile Memory

The quest for universal memory featuring high density, fast operation, and low power involves a large diversity of memory candidates such as MRAM (Magnetic RAM), PCRAM (Phase Change RAM), Fe-RAM (ferroelectric RAM), and molecular memories. Recently the ferroelectricity of some ultrathin films has been shown to be a good candidate for nonvolatile memory thanks to the bistability of polarization [13–19]. In ferroelectric materials, the electrical dipoles orientate according to an applied external field, and, after reaching a critical electric field, the resulting orientation can be maintained even when the field is switched off. This mechanisms can be used to store charges and therefore to store information for nonvolatile memory applications.

Standard ferroelectric materials like strontium–barium–titanate (SBT) and lead–zirconium–titanate (PZT), employed in some commercial Fe-RAM, still have problems in terms of film quality and retention time [20, 21]. Moreover these perovskite ferroelectrics require high-temperature annealing, which makes them incompatible with CMOS coprocessing. Recently, vinylidene fluoride (VDF) and copolymer with trifluoroethylene (Tr-FE) have been attracting growing interest because of their large spontaneous polarization ($\sim$0.1 C/m^2), polarization stability, switch time ($<$0.1 μs), and suitability for organic devices [17]. Several P(VDF-TrFE) bistable capacitors have been reported with high operating voltages (tens of volts) and good retention times, while the very few low-operating-voltage devices showed reduced low retention ($\sim$15 min). In this work, we concentrate our efforts on the processing and study of the properties of ultrathin P(VDF-TrFE) films when integrated in the gate stack of a MOSFET, which could operate as a one-transistor (1T) memory cell as they would be more scalable than conventional 1T–1C memory architectures using ferroelectric capacitors.

Fe-FET Fabrication

The proposed process targets the use of sub-100-nm ultra-thin-layer PVDF copolymers in the gate stack of *n*-channel MOSFET on bulk silicon. Device active areas and STI isolation are UV-lithographically defined on *p*-doped silicon. Two different substrates have been used: a low-resistivity ($N_A = 4 \times 10^{16}$ cm^{-3}) wafer for 40 nm P(VDF-TrFE) and a high-resistivity one for 100 nm P(VDF-TrFE) ($N_A = 10^{15}$ cm^{-3}). Source and drain regions are highly doped (10^{20} cm^{-3}) and 10 nm of SiO$_2$ is thermally grown. The P(VDF-TrFE) (70–30%) is prepared using an original recipe, similar to a recent report [22] based on methyl-ethyl-ketone, optimized to considerably reduce the film thickness (into the sub-100-nm range) and, consequently, the voltage for coercive field. The solution is spin-coated and baked for 5 min at 137°C. Two polymer layer thicknesses, of 100 and 40 nm, are studied.

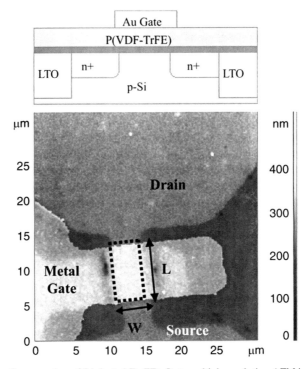

Fig. 2.11 *Top*: Cross-section of fabricated Fe-FEt. *Bottom*: high-resolution AFM image of PVDF-TrFE gate stack of fabricated Fe-FET. The average roughness is 38 nm (which limits the realization of films thinner than 40 nm)

Atomic force microscope (AFM) measurements have been systematically used to evaluate the topography of the copolymer films and optimize their preparation. A thin gold layer (100 nm) defines the gate contact. MOSFET devices with different channel lengths, L, and widths, W, ranging from 2 to 50 μm, have been designed; a fabricated device is shown in Fig. 2.11.

Experimental Results

The measurements of the polarization and of the capacitance on the gate stack of the fabricated Fe-FET prove that the polymer has been successfully integrated onto a silicon substrate. In fact, the hysteretic loop indicates that the ferroelectric properties of the material are preserved following the fabrication process. Figure 2.12 shows the polarization for two different samples: the first with only a P(VDF-TrFE) gate layer and the second with a SiO_2/P(VDF-TrFE) gate stack. Both samples have a ferroelectric layer of 100 nm. The different bending of the curves is due to the drop in voltage on the SiO_2 layer in the second sample. The P(VDF-TrFE) polymer has

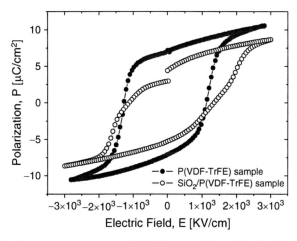

Fig. 2.12 Polarization measurements on two Fe-FET gate stacks: with and without SiO2 layer for a device with $L = W = 50\,\mu m$

a remanent polarization (Pr) of about $8\,\mu C/cm^2$ and a coercive field (E_c) of about $1.3\,MV/cm$.

At first view, the output characteristics, I_D–V_D, of the Fe-FET show linear and saturation regions qualitatively comparable with those of a constant gate oxide capacitance MOSFET (Fig. 2.13). However, there is a combined effect of V_G and V_D on the gate polarization (and consequently on the value of the gate-to-channel capacitance), which influences the buildup of inversion charge and the saturation voltage V_{Dsat} in a more complex manner than the established equations for a standard MOSFET, which also depend on device size and geometry.

The I_D–V_G electrical characteristics are evaluated for both 100- and 40-nm ferro-electric layers, showing that, like conventional MOSFET, the gate voltage controls the drain current. An equivalent threshold voltage exists, with a very abrupt transition between the off and on states and an associated hysteresis. The memory effect needed for the 1T-cell is clearly demonstrated by the hysteretic loops reported at room temperature in Fig. 2.13a, b. In the case of 40-nm PVDF-TrFE the I_{on}/I_{off} is of the order of 10^5 to 10^6 (depending on the applied V_D) and the off-state current is less than 10^{-12} A. We explain the saturation of the drain current by the saturation of the polarization.

The saturation of the drain current is reached at around 10 V for the 40-nm dielectric (Fig. 2.4), less than half of the voltage needed for a thickness of 100 nm. These values are in good agreement with theory relating the voltage to the coercive field (E_c) and ferroelectric dielectric thickness; in fact the saturation voltage V_{sat} is given by $V_{sat} = E_c d$, where d is the layer thickness and E_c is the coercive field. For the 100-nm thickness we found a V_{sat} of about 15 V, and for the 40-nm case its value drops to 7 to 8 V, these theoretical values fit well the experimental ones. Moreover, from these results it is observed that the coercive field E_c is almost the same for both 100- and 40-nm thicknesses.

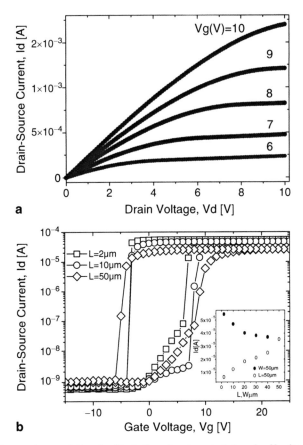

Fig. 2.13 **a** Output characteristics, I_D–V_D. **b** Transfer characteristics, I_D–V_G, for a Fe-FET with 100-nm P(VF-TrFE) in gate stack and $L = W = 50\,\mu m$

The comparison of the I_D–V_G characteristics of 40- and 100-nm PVDF-TrFE transistors, with the same design ($L = W = 10\,\mu m$), demonstrates significant improvement in terms of memory operating voltage (hysteresis) for the 40-nm polymer thickness (Fig. 2.14). While the I_{off} current difference (cf. $I_{off}\sim10^{-9}$ A, in Fig. 2.13b with $I_{off}\sim10^{-11}$ A, in Fig. 2.14, for the same device width) is mainly due to the difference in the substrate doping levels; the I_{on} reduction for the 40-nm thickness is explained by a more diluted solution that results in a ferroelectric polymer material with less polarization.

The Fe-FET has also been characterized in terms of memory properties: retention time, endurance, and programming time [23]. The experiments show an $I_D(V_G)$ hysteresis fairly symmetric with respect to 0 V, enabling an efficient read operation. In fact the binary values (0 or 1) stored in the 1T Fe-FET cell can be read out at $V_G = 0$ V (Fig. 2.15). Retention analysis has been performed by applying to the gate a 20-V pulse signal with a width of 30 s and measuring the drain current. Figure 2.15

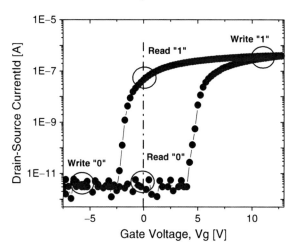

Fig. 2.14 I_D–V_G characteristics and programming regions for a Fe-FET device operated as 1T memory cell

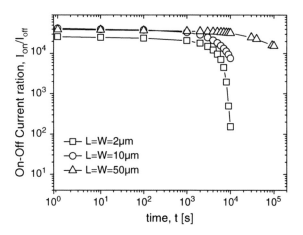

Fig. 2.15 Retention time of 1T Fe-FET memory cell with 100-nm polymer thickness, after writing the high state "1" with $V_g = 20$ V for 30 s, at room temperature

shows the change in the I_{on}/I_{off} ratio as a function of time. The largest Fe-FET ($L = W = 50\,\mu$m) has a retention time of about 2 d, while the smallest one ($L = W = 2\,\mu$m) of about 3 h; retention time dependence on device size is under investigation. This dependence has already been observed in PZT/SBT ferroelectric capacitors. In a number of publications, the switching polarization loss was attributed to the lateral damage provoked by etching [18] or electric field nonuniformity at the Fe-CAP edges [24]. A preferential polarization state observed in very small Fe-CAPs was explained by the accumulation of defects at the lateral surface [25]. In most recent works it was demonstrated that the size effects consist not only of a

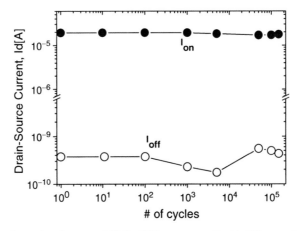

Fig. 2.16 Experimental endurance of 1T Fe-FET memory cell with 100-nm polymer thickness; the gate is cycled with 20Vpp square signal, and I_{on} and I_{off} are evaluated after each decade of cycles

decrease but also of a redistribution and even inversion of the original spontaneous polarization [26].

Finally, typical 1T memory endurance experimental analysis has been performed and is reported in Fig. 2.16; after 10^5 cycles, both I_{on} and I_{off} current levels are well preserved, suggesting the suitability of the investigated new material for nonvolatile memory cells.

Discussion

We have integrated for the first time a P(VDF-TrFE) copolymer into the gate stack of a standard MOSFET structure and demonstrated a fully operational 1T nonvolatile memory cell. Devices with two different copolymer thicknesses (40 and 100 nm) have been characterized and a memory window of 6 V has been found for the thinner dielectric. The I_{on}/I_{off} ratio of fabricated Fe-FETs is in the order of 10^5, with I_{off} currents ranging from 10^{-11} to 10^{-9} A. The Fe-FET memory cell has a retention time of about a few days, a good endurance (10^5 cycles demonstrated), and a programming time on the order of milliseconds (due to the micrometer size of the fabricated devices). These performances make the 1T Fe-FET memory cell suitable for any nonvolatile memory application requiring a storage time of some days, with an associated low cost. Future work is foreseen to improve the retention characteristics of the reported memory cell and to develop accurate modeling for the Fe-FET electrical operation.

Materials for Piezoelectric Nanodevices

Recent examples of piezoelectric materials in polymer matrix show that sensors have established a solid presence in our daily lives from ultrasound applications in medicine, to underwater ultrasound in military and civilian applications, to smart sensor systems in automobiles, or to nondestructive testing in industry. Since the 1950s, lead titanate ($PbTiO_3$) compounds have been the leading materials for piezoelectric applications due to their excellent dielectric and piezoelectric properties. The rugged solid-state construction of industrial piezoelectric ceramic sensors enables them to operate under most harsh environmental conditions. However, the production of piezoelectric nanowires and their applications are still very challenging.

With the increasing public awareness of health concerns associated with lead and the recent changes in environmental policies, we have focused our research on $KNbO_3$-based nanowires. Alkali niobate materials with a perovskite-type structure are believed to be the best candidates to replace piezoelectric lead-containing materials. On the other hand, $KNbO_3$ is a photocatalyst with E_g of about 3.3 eV. $KNbO_3$ nanowire arrays could be suitable materials to compete with TiO_2 for the production of H_2 by water splitting in a new generation of photocatalytic cells.

Composite $KNbO_3$ Nanowires: Epoxy Resin for Piezoelectric Devices

Single-crystal $KNbO_3$ nanowires are grown by hydrothermal treatment from a mixture of Nb_2O_5 and highly concentrated KOH solution [27]. After optimizing the growth conditions [27] and studying the kinetics of the chemical reactions [28], we have modified the method to produce a 1×1 cm carpet of $KNbO_3$ nanowires epitaxially grown on a [001] $SrTiO_3$ substrate [29] (Fig. 2.17). As previously reported (see 2007 CCMX interim report), the morphological characteristics (length, diameter, etc.) and density of nanowires strongly depend on the roughness of the $SrTiO_3$ substrates. It is believed that structural defects induced by the polishing process on the substrate surface act as nucleation sites for the crystallization of the $KNbO_3$ nanowires.

The nanowire arrays are subsequently embedded in a polymer matrix by immersion in an epoxy resin and the composite is easily peeled off from the $SrTiO_3$ substrate. Excess polymer is removed by gentle polishing. Gold is subsequently evaporated on both sides of the composites. Innumerous samples, with approximate dimensions of $500 \times 500\,\mu m$ and $50\mu m$ thickness, were prepared and the piezoactivity was measured by placing the sample inside a pressure cell that was connected to an electrometer measuring charge, and hydrostatic pressure up to 3.5 kbar was applied. The signal of a strain gauge measuring the applied force and

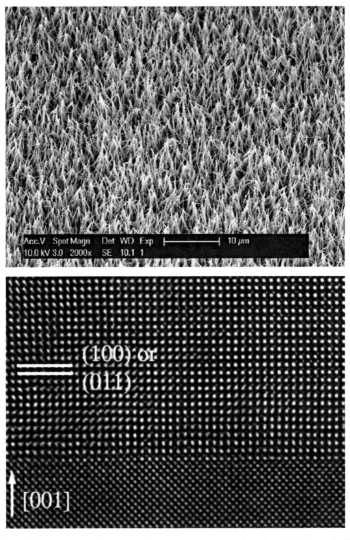

Fig. 2.17 SEM images of KNbO$_3$ nanowires carpet grown on SrTiO$_3$ substrates and HRTEM of the SrTiO$_3$–KNbO$_3$ interface

the analog output of the electrometer were recorded with an oscilloscope. Several charge-discharge curves were measured (Fig. 2.18).

Although these results are very encouraging, the measured piezoactivity of the composite differs significantly from the value calculated from the bulk KNbO$_3$ piezoelectric coefficient, which is on the order of 10 pC/N [30]. Actually, the KNbO$_3$ nanowires are multidomain with a random polarization direction. Additional investigations are being carried out while applying an electric field to polarize entirely the nanowire along their axis prior to the application of pressure.

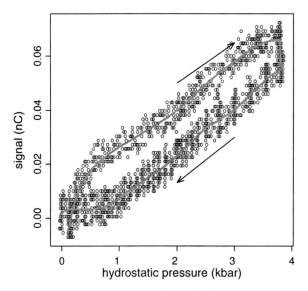

Fig. 2.18 Piezoelectric hysteresis loop obtained from KNbO$_3$ nanowires – epoxy composites. *Arrows* indicate direction of scan. Original data points are indicated by *symbols*, *solid line* is a guide for the eye

Water Splitting Photocatalyzed by $KNbO_3$ Nanowires: Discussion

In order to investigate the photocatalytic activity of KNbO$_3$ nanowires, we have implemented an electron spin resonance (ESR) technique to detect and quantify singlet-oxygen generation. Regarding the photosynthesis process, H$_2$ is formed via light-stimulated exciton generation in the bulk of the KNbO$_3$ nanowires. Excitons can be understood as e–h pairs. The extra, light-induced electrons in the conduction band are needed to initiate the H$_2$ formation process at the particle surface. Therefore, the evidence of singlet-oxygen generation will be closely related to the effective exciton formation under stimulation with UVA. Thus, it gives a good prognostic for H$_2$ formation.

The preliminary results of our ESR investigations are plotted in Fig. 2.19. The linear evolution of the ESR signal intensity vs. time shows the ability of KNbO$_3$ nanowires to generate singlet oxygen. However, this production is still reduced compared to TiO$_2$ films.

As perspectives, the photocatalytic activity will be tentatively enhanced by decorating the KNbO$_3$ nanowire surface with cocatalyst Ni/NiO particles, which are known to work as an H$_2$ evolution site. In addition, the ESR analysis will be combined with mass spectrometry to substantiate the correlation between singlet-oxygen generation and H$_2$ evolution.

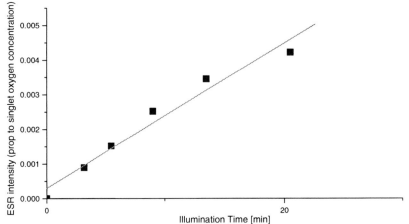

Fig. 2.19 Evolution of ESR signal intensity, proportional to the singlet oxygen concentration generated by illumination of the KNbO$_3$ nanowires with UVA

Conclusions

Various materials have been shown to be very relevant for the construction of nano-electronic circuits. The synergy between material choice and electronic design is extremely important especially in view of achieving the desirable electrical and mechanical properties. To date, a large number of materials have been used in micro- and nanoelectronics, but a broad range of opportunities remain untapped.

References

1. Dennard RH, Cai J, Kumar A (2007) A perspective on today's scaling challenges and possible future directions. Solid State Electron 51(5):18–25
2. Colinge J-P (1997) Silicon-on-insulator technology: materials to VLSI, 2nd edn. Kluwer, Dordrecht
3. Moselund KE (2008) Three-dimensional electronic devices fabricated on a top-down silicon nanowire platform. EPFL PhD Thesis
4. Moselund KE, Bouvet D, Leblebici Y, Ben Jamaa H, Atienza D, De Micheli G, Ionescu AM (2007) GAA MOSFET for logic on a wire. In: European Nanoday. International nanotech-nology conference on communication and cooperation, INC3 2007, Brussels, Belgium, 16–19 April 2007
5. Moselund KE, Pott V, Bouvet D, Ionescu AM (2007) Abrupt current switching due to impact ionization effects in Ω-gate MOSFET on low doped bulk silicon. ESSDERC 2007, Technical Digest, pp 287–290
6. Moselund KE, Bouvet D, Ionescu AM (2007) Prospect for logic-on-a-wire: Ω-gate NMOS inverter fabricated on single Si nanowire. Technical Digest MNE 2007, Copenhagen, 5B-2, pp 123–124

7. Moselund KE, Dobrosz P, Olsen S, Pott V, De Michielis L, Tsamados D, Bouvet D, O'Neill A, Ionescu AM (2007) Bended gate-all-around nanowire MOSFET:a device with enhanced carrier mobility due to oxidation-induced tensile stress. Technical Digest IEDM, pp 191–194

8. Moselund KE, Bouvet D, Ben Jamaa MH, Atienza D, Leblebici Y, De Micheli G, Ionescu AM (2008) Prospects for logic-on-a-wire. Microelectronic Eng (Special Edition) 85:1406–1409

9. Moselund KE, Pott V, Bouvet D, Ionescu AM (2008) Hysteretic inverter-on-a-body-tied-wire based on less-than-10 mV/decade abrupt punch-through impact ionization MOS PIMOS switch. VLSI-TSA, Taiwan, April 2008

10. Moselund KE, Pott V, Meinen C, Bouvet D, Kayal M, Ionescu AM (2008) DRAM based on hysteresis in impact ionization single-transistor-latch. Accepted for presentation at the MRS Spring meeting, San Francisco, March 2008

11. Moselund KE, Bouvet D, Pott V, Meinen C, Kayal M, Ionescu AM (2008) Punch-through impact ionization MOSFET (PIMOS):from device principle to applications. Solid State Electron 52:1336–1344

12. Gopalakrishnan K, Griffin PB, Plummer JD (2005) Impact ionization MOS (I-MOS) – Part I: Device and circuit simulations. IEEE Trans Electron Devices 52(1):69–76

13. Furukawa T, Date M, Ohuchi M, Chiba A (1984) Ferroelectric switching characteristics in a copolymer of vinylidene fluoride and trifluorethylene. J Appl Phys 56(5):1481

14. Lim SH, Rastogi AC, Desu SB (2004) Electrical properties of metal-ferroelectric-insulator-semiconductor structures based on ferroelectric polyvinylidene fluoride copolymer film gate for nonvolatile random access memory application. J Appl Phys 96(10):5673

15. Gerber A, Kohlstedt H, Fitsilis M, Waser R, Reece TJ, Ducharme S, Rije E (2006) Low-voltage operation of metal-ferroelectric-insulator-semiconductor diodes incorporating a ferroelectric polyvinylidene fluoride copolymer Langmuir-Blodgett film. J Appl Phys 100:024110

16. Reece TJ, Ducharme S, Sorokin AV, Poulsen M (2003) Nonvolatile memory element based on a ferroelectric polymer Langmuir–Blodgett film. Appl Phys Lett 82:142

17. Naber RCG, Tanase C, Blom PWM, Gelinck GH, Marsman AW, Touwslager FJ, Setayesh S, Leeuw DM (2005) High-performance solution-processed polymer ferroelectric field effect transistors. Nat Mater 4(3):243

18. Sakoda T et al (2001) Hydrogen-robust submicron $IrO_x/Pb(Zr, Ti)O_3/Ir$ capacitors for embedded ferroelectric memory. Jpn J Appl Phys 40(1):2911

19. Kinam K, Jung Hyuk C, Jungdal C, Hong-Sik J (2005) The future prospect of nonvolatile memory. 2005 IEEE symposium on VLSI technology

20. A-Paz de Araujo C, Cuchiaro JD, McMillan LD, Scott MC, Scott JF (1995) Fatigue-free ferroelectric capacitors with platinum electrodes. Nature 374(6523):627

21. Nakao K, Judai Y, Azuma M, Shimada Y, Otsuki T (1998) Voltage shift effect on retention failure in ferroelectric memories. Jpn J Appl Phys 37:5203–5206

22. Fujisaki S, Ishiwara H, Fujisaki Y (2007) Low-voltage operation of ferroelectric poly(vinylidene fluoride-trifluoroethylene) copolymer capacitors and metal-ferroelectric-insulator-semiconductor diodes. Appl Phys Lett 90:158

23. Salvatore GA, Bouvet D, Stolitchnov I, Setter N, Ionescu AM (2008) Low voltage ferroelectric FET with sub-100 nm copolymer P(VDF-TrFE) gate dielectric for non-volatile 1T memory. ESSDERC 2008, Edinburgh, Scotland, 15–19 September 2008

24. Paz de Araujo CA, Cuchiaro JD, McMillan LD, Scott MC, Scott JF (1995) Fatigue-free ferroelectric capacitors with platinum electrodes. Nature 374(6323):627–629

25. Alexe M, Harnagea C, Hesse D, Gösele U (2001) Polarization imprint and size effects in mesoscopic ferroelectric structures. Appl Phys Lett 79:242

26. Stolitchnov I, Colla E, Tagantsev A, Bharadwaja SSN, Hong S, Setter N, Cross JS, Tsukada M (2002) Unusual size effect on the polarization patterns in micron-size Pb(Zr,Ti)O-3 film capacitors. Appl Phys Lett 80(25):4804–4806

27. Magrez A, Vasco E, Seo JW, Dieker C, Setter N, Forro L (2006) .Growth of single-crystalline KNbO3 nanostructures. J Phys Chem 110:58–61

28. Vasco E, Magrez A, Forro L, Setter N (2005) Growth kinetics of one-dimensional KNbO3 nanostructures by hydrothermal processing routes. J Phys Chem B 109(30):14331–14334
29. Magrez A, Seo JW, Dieker C, Forró L Preparation of large scale arrays of oriented KNbO3 nanowires, in preparation
30. Wada S, Muraoka K, Kakemoto H, Tsumuri T, Kumagai H (2005) Enhanced piezoelectric properties of potassium niobate single crystals with fine engineered domain configurations. Mater Sci Eng B 120:186–189

Chapter 3
Design Technologies for Nanoelectronic Systems Beyond Ultimately Scaled CMOS

**Haykel Ben Jamaa, Bahman Kheradmand Boroujeni,
Giovanni De Micheli, Yusuf Leblebici, Christian Piguet,
Alexandre Schmid, and Milos Stanisavljevic**

Introduction

As already explained in the introduction to Chap., the development of economically feasible nanoelectronic systems requires a tight interplay between materials and fabrication technologies on the one hand and design technologies on the other. In particular, it is quite essential to explore *circuit-level measures* to mitigate the limitations of process variations (PVs), leakage, and reduced device reliability and, finally, to explore *system-level design* approaches that are better adapted to the constraints imposed by the materials, technology, and device physics. This chapter largely deals with some of these key questions that relate to design technologies for nanoelectronic systems.

Fault-tolerant design approaches for regular arrays based on silicon nanowires are discussed in detail in the section entitled "Fault-tolerant Design Approaches for Regular Arrays Based on Silicon Nanowires." Turning the focus more toward conventional technologies, the next section explores a novel technique for minimizing local delay variations for nanoscale CMOS technologies. Finally, the last section presents the adaptive V_{gs} technique for controlling the power and delay of nanometer-scale logic gates operating in a sub-VT regime.

Fault-Tolerant Design Approaches for Regular Arrays Based on Silicon Nanowires

With the progress of manufacturing technologies, many one- and zero-dimensional electronic devices have been designed and their operation demonstrated. These devices, which include nanowires (NWs) [1] and molecular switches [2], promise an ultrahigh integration density and push the fabricated circuits a few steps closer to

Yusuf Leblebici (✉)
EPFL – Swiss Federal Institute of Technology, Lausanne, Switzerland

G. De Micheli et al., *Nanosystems Design and Technology*,
DOI 10.1007/978-1-4419-0255-9_3, © Springer Science+Business Media LLC 2009

the natural limits imposed by the physics of electron-based systems. In their mature stage, they did not reach such a level that their placement could be controlled in an accurate and cost-effective way. One promising design paradigm is based on a regular arrangement of these devices into arrays [3], which offers an easier fabrication approach and a higher fault tolerance and design flexibility thanks to the inherent redundancy level.

In this section we investigate different design approaches for regular arrays based on silicon NWs. We first present a CMOS-compatible fabrication technique for regular silicon NW arrays. Then we present some design challenges for regular NW circuits based on the example of crossbar memories. The challenges covered here include decoder design and reliable circuit testing.

Fabrication of Nanowire Arrays

A crossbar circuit is formed by two perpendicular layers of parallel NWs with molecular switches at their cross-points (Fig. 3.1). These molecular switches can store information (for memories) or perform computation (for logic).

It is of considerable interest to build arrays of dense parallel NWs while meeting the following requirements. On the one hand, it is desirable that the fabrication process be compatible with standard CMOS processes, not only for cost reasons, but also in order to integrate crossbar circuits onto CMOS chips. On the other hand, the NWs, which have a sublithographic resolution, need to be contacted by the lithographically defined outer circuit.

We propose the use of the spacer patterning technique (SPT) in order to address these issues. This technique was successfully used to build FinFET with sublithographic dimensions [4] by transforming a thin vertical dimension into a narrow horizontal dimension. An iteration of this technique [5] allows the fabrication of dense arrays of a few nanometer-wide parallel wires (Fig. 3.2).

The addressing of the NWs can be performed by a lithographically defined single-gate electrode laid out over the whole NW set. The current flow in the NWs is field-effect-controlled. Assuming that the NW FETs have different threshold voltages, one single-gate electrode is able to modulate the current flow through the

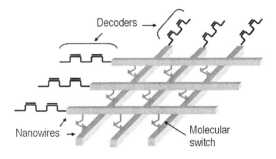

Fig. 3.1 Overall organization of a crossbar circuit

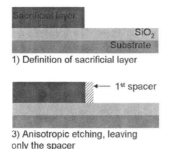

1) Definition of sacrificial layer

2) Conformal deposition

3) Anisotropic etching, leaving
only the spacer

4) Iteration of steps 2-3 with
different materials

Fig. 3.2 Multispacer patterning technique

Fig. 3.3 SEM cross-
section of multispacer
$(3 \times poly - Si + 2 \times SiO_2)$

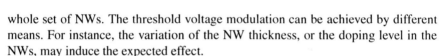

whole set of NWs. The threshold voltage modulation can be achieved by different
means. For instance, the variation of the NW thickness, or the doping level in the
NWs, may induce the expected effect.

We applied this approach to fabricate arrays of long NWs and to address them.
In Fig. 3.3 we show some scanning electron microscopy (SEM) pictures of the array
cross-section with three to five parallel spacers.

Addressing Nanowire Arrays

The pitch of NW arrays is not dependent on the (photo-) lithography limit. For
instance, the fabricated array shown above has a pitch of 150 down to 25 nm, and
it is below the lithography pitch, as drawn in the layout. It is therefore necessary to
address the issue of contacting and addressing every NW independently of the others
in the same array. In what follows, we investigate the design aspects of the decoder
that guarantees a unique addressing of NWs under high variability conditions. In
order to investigate the design space with a concrete circuit, we assumed that the
NW array operated as a crossbar memory.

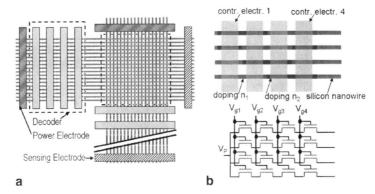

Fig. 3.4 **a** Crossbar memory with decoder. **b** Decoder layout and circuit

Nanowire Array Model and Technology

The crossbar memory architecture is depicted in Fig. 3.4a. The studied architecture has two parts, organized in an identical way and laid out perpendicularly to each other. Each part is a plane of N parallel NWs. M mesowires are used to address the NWs within each group. Then the NW decoder (Fig. 3.4b) has the size $N \times M$. The area sandwiched between the NW arrays is the actual memory, in which the information is stored in bistable switches grafted at the cross-points.

Each NW has to be addressed. This operation is performed by the NW decoder, which is formed by a set of parallel mesowires (or control wires) crossing the NW plane. The part of NWs under the decoder is coated by a dielectric, thus allowing a field-effect control of the NWs by the mesowires of the decoder. The NW technology assumed for this decoder exploits the bulk silicon fabrication platform reported in [6]. The proposed access devices in the decoder are gate-all-around (GAA) FETs whose threshold voltage depends on the doping level of the channel. We considered the use of a multiple-threshold voltage process that enables the fabrication of a multivalued logic decoder.

Multivalued Logic Codes for Nanowire Arrays

Each NW has a series of differently doped regions, defining a multivalued logic pattern. Each applied address is a series of voltages defining a multivalued logic code that covers a pattern, switches on all the transistors in the NW, and lets the current flow through it. A NW is said to be uniquely addressed by a code if this code covers only this NW pattern.

We can generalize the notion of binary code to multivalued logic by defining two types of codes: the n-ary hot code and the n-ary reflexive code. Both of them are addressable, i.e., a NW array coded with either an n-ary hot or reflexive code has every NW addressed uniquely for any applied address. However, when defects affect the array, the addressing becomes more complex, as explained in what follows.

The control of silicon NWs is based on the modulation of the threshold voltage of the controlling transistors. The encoding schemes impose a distribution of the applied control voltages between the successive threshold voltages (V_T). The main issue with V_T is its variability and process-dependency. It is generally assumed that it follows a normal distribution.

We define a single-digit error as follows. If V_T exceeds a certain value, then the corresponding digit with value i will be detected as $(i + 1)$, and this is called a flip-up defect. The complementary case is when the threshold voltage drops beyond a fixed value, and then digit i is detected as $(i - 1)$, and this is called a flip-down defect.

If V_T varies within a small range close to its mean value, then the pattern does not change since the NW still conducts under the same conditions. Then, a one-to-one1 mapping between the code and the pattern space holds. By contrast, if the V_T variation is large, then some digits may be shifted up or down, as explained above. When a pattern has a sequence of errors, either it can be covered by one or more codes or it can be covered by no code. When we consider the codes, some of them cover one or more patterns and others cover no pattern under the error assumptions.

We defined the algorithms that estimate for every code type and length the part of NW patterns that become nonaddressable under given variability conditions [8]. The results could be confirmed by Monte Carlo simulations, which enable a more accurate assessment of the code behavior and a better exploration of the decoder design space.

Design of Nanowire Array Decoders

We considered a crossbar memory based on a double layer of NW arrays including the decoders, and we estimated the effective memory capacity in terms of defect-free and addressable bits per unit area. The effective memory capacity for different codes and technologies is depicted in Fig. 3.9a and shows an area saving up to ~20% depending on the decoder design choices. Figure 3.5b shows the impact of the variation in the applied voltages at the decoder from its nominal value (v) for a high variability level; the figure demonstrates that an optimized choice of the applied voltages improves the effective memory capacity.

Testing Nanowire Memories

Even though there are no complete memory systems based on the crossbar architecture yet, we believe that such systems will have the same architecture as CMOS memories [7, 9] (Fig. 3.6a). Unlike conventional RAM, crossbar memories have

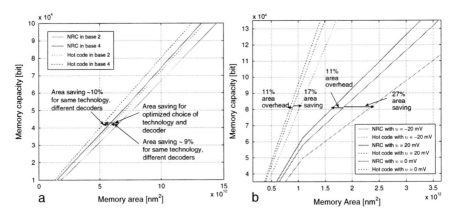

Fig. 3.5 **a** Area-capacity tradeoff for different codes. **b** Area-capacity tradeoff for different applied voltages

Fig. 3.6 Crossbar system

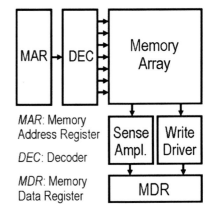

two parts: a sublithographic part formed by the decoder and the memory array and fabricated using one of the emerging technologies described in the section titled "Fabrication of Nanowire Arrays" and a lithographic part formed by the rest of the circuit and fabricated using CMOS technology.

The information is assumed to be stored in molecular switches grafted to every pair of crossing NWs. In the on-state, the molecule is conducting (logic 1), and in the off-state, it is highly resistive (logic 0). The writing operation is performed by first selecting the bit to be written, and then by applying a large positive or negative voltage at the pair of NWs connected by the molecular switch in order to set the molecular state, i.e., the bit value. On the other hand, the reading operation is current-based. In fact, if the molecule is in the off-state, then the NW in the lower level is almost floating and no correct voltage level can be sensed. Consequently, the reading operation is performed by selecting the bit to be read, then by measuring

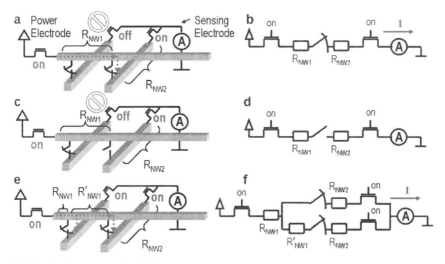

Fig. 3.7 Read operation in a 2-bit memory

the current through the sensing electrode (Fig. 3.7a–d). Thus, the current-based read operation in crossbar memories necessitates a thresholder as a part of the sense amplifier in order to set the limit between the logic values 0 and 1 and to translate them into logic levels that can be stored in the memory data register.

Variability-Induced Errors in Crossbar Memories

The threshold voltage variation was shown to cause defects in the decoder in such a way that by applying an address, any number of NWs could be activated instead of one single NW. Figure 3.7e, f shows an example of defective addressing in the second NW layer. Thus, the sense amplifier reads the superposition of the information stored in two bits. The thresholder cannot properly distinguish between the sensed signals resulting from the following cases: (a) one bit with a value of 1 and (b) the superposition of two bits whereby at least one of them has a value of 1. In such a situation, the read operation of the first bit yields a result depending on the state of the second bit, which causes coupling faults (CF) in the memory [7]. Considering the fact that decoder defects typically make two, three, or more NWs in each array active with the same address [8], the number of interdependant bits can be as large as 4 to 9 or even more, without necessarily having neighboring locations. This leads to the more critical pattern sensitivity faults [7].

In order to avoid complex and exhaustive PSF test procedures on the whole memory [9], one may try to resolve the PSF caused by the decoder defects before performing the conventional memory test. The thresholder can carry out this operation by checking the addresses of all NWs in every layer (after separating them) and keeping only the addresses that activate one single NW. This procedure has a linear complexity with N, the number of NWs in a layer (where N^2 is the number of bits in the memory). While it represents an additional testing step, this testing procedure, which we call a *nanowire test*, obviates the necessity of an exhaustive PSF testing of the whole memory whose complexity is exponential with N^2. However, we expect that the molecular switches will also induce PSFs that we do not consider in this paper. Since only neighboring molecules are likely to interact with each other, one can assume neighborhood patterns for the PSFs caused by the molecules. Therefore, simplified PSF procedures having a linear complexity with N^2 can be applied [9].

Thresholder Optimization

For every address applied at the decoder, a validation signal is given by the thresholder indicating whether (a) a single NW is addressed or (b) no NW or more than one NW is addressed. The thresholder senses I_s, after a possible amplification, then it compares it to two reference values (I_0 and I_1 with $I_0 < I_1$). If the sensed current is smaller than I_0, then no NW is addressed. If the sensed current is larger than I_1, then at least two NWs are activated with the same address. If the sensed current is between the reference current levels, then only one NW is activated and the address is considered to be valid.

In our defect model we assume two sources of variability of the sensed current: (a) variation of the threshold voltages of the transistors in the decoder: we assume, as for conventional MOS devices [4], that V_T follows a normal distribution with known mean value V_T and standard deviation σ_T; (b) variation of the NW resistance in the memory array: in the mathematical model, we assume that this resistance is fixed. Then, we investigate the impact of its variation on the results.

By assuming that the V_Ts are stochastic variables and the NW resistance is fixed, we model I_s as a stochastic variable whose distribution parameters depend on the NW resistance. Thus, the calculation of the thresholder parameters I_0 and I_1 results from a stochastic optimization. Their optimal values are obtained by maximizing the probability that a correct address is detected (P_1: the conditional probability that I_s is between I_0 and I_1 given that only one NW is activated) and the probabilities that a defective address is identified as such (P_0 and P_2: the conditional probability that I_s is below I_0 or beyond I_1 given that no NW or more than one single NW is activated, respectively). Then, the probability that all three events happen simultaneously is given by $P_0 \times P_1 \times P_2$ (assuming that the considered events are independent). Consequently, we can define the error probability of the thresholder as $\varepsilon = 1 - P_0 \times P_1 \times P_2$. In order to optimize the design of the thresholder with the smallest error, we developed a model that calculates the optimal values of I_0 and I_1.

Fig. 3.8 Circuit model for a NW under test

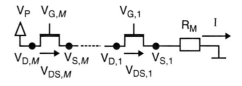

Stochastic and Perturbative Model

It is possible from a mathematical point of view to derive the exact stochastic distribution of the sensed current as a function of the stochastic distributions of the threshold voltages. However, the nonlinear effects, especially those affecting the short channel transistors, make an analytical solution impossible. Nevertheless, the model can be kept simple by linearizing the decoder circuit around the operating point. Each addressed NW was modeled as shown in Fig. 3.8. From this circuit model, the linearized relations between current and voltages can be derived.

On the other hand, the sensed current (either before or after linearization) can be modeled with different components. The first component is the useful signal, which is the current flowing through a correctly (i.e., intentionally) addressed NW. The second component is the defect-induced signal, which is represented by the current flowing through a badly (i.e., unintentionally) addressed NW. The third component is the intrinsic noise, which is the (subthreshold) current flowing through nonaddressed NWs.

We quantified these three stochastical components of the current and linearized them around the operating point. The obtained linearized stochastical current model was used in order to optimize the thresholder parameters I_0 and I_1 [10].

Thresholder Design

The design of the thresholder means in this context the optimization of its parameters I_0 and I_1 and the minimization of the error ε. The optimal value of I_1 is shown for different design and technology parameters in Fig. 3.9a. It is fairly robust and constant for a low memory resistive load. The value $I_0 = 0.8 \times I_1$ was found to be a good compromise between the cancellation of intrinsic noise and a signal that is useful for detection. The error of the decoder was assessed in Fig. 3.9b for different design and technology parameters. Adding redundancy to the decoder by increasing the code length is shown to be a very efficient way to minimize thresholder error. For instance, using 18 instead of 12 access transistors reduces the error by a factor of 60.

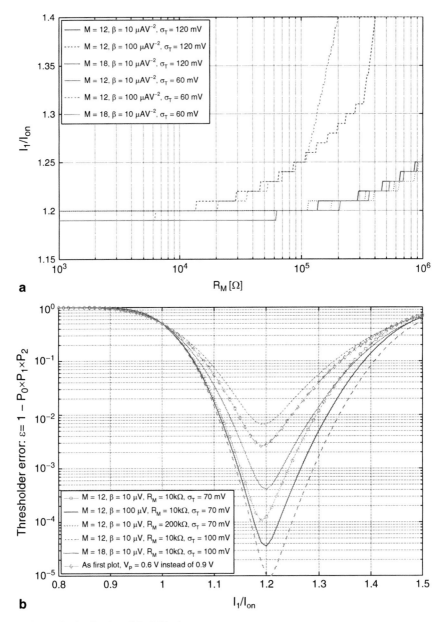

Fig. 3.9 **a** Optimal value of I_1. **b** Testing error

Summary

This section has shown the promises of crossbars when used as regular macros for computation and storage. The compact nature of crossbars leads to an efficient use of silicon areas. Nevertheless, the connection of these nanoelements to mesowires requires specific addressing and decoding techniques. A complete design style has been described, including a procedure for testing.

Minimizing Local Delay Variations for Nanoscale CMOS Technologies

The technology scaling that has been the trend for decades is expected to continue at the same speed or possibly at a slightly slower pace for at least the next 10 years. The nano age has already begun (where typical feature dimensions are considered to be less than 100 nm). According to the ITRS roadmap, the operation frequency is expected to increase up to 12 GHz and a single chip could contain over 12 billion transistors in 2020 [11]. Future very-deep submicron and nanoelectronic fabrication technologies are expected to suffer from the dramatic dimensional scaling, which will strongly impact parameter variations.

The yield of low-voltage digital circuits is found to be sensitive to die-to-die (D2D) (interdie, global), and within-die (WID) (intradie, local) parameter variations in the manufacturing process. D2D variations act globally on the entire chip or on functional blocks, so that each device on one chip or in one block shows the same deviation. Interchip or interblock variations can be caused by systematic effects like process gradients over the wafer [12] with typical distances in the range of functional block sizes or above. Variations of the gate oxide thickness can be regarded as global variations. Sets of worst-case and best-case parameters are used during design verifications to mitigate the impact of global variations. The effects of WID variations are becoming more and more prominent with scaling and they have a direct influence on local gate delay variations. Numerous random factors such as statistical deviations of the doping concentration and imprecision of lithography lead to more pronounced delay variations for minimum transistor sizes [13, 14]. These factors are intrinsic since they cannot be eliminated by external control of conventional manufacturing. The increase of the path delay variations for smaller device dimensions and reduced supply voltages, as well as the dependence of path delay variations on the path length, are becoming more prominent with scaling. Circuits with a large number of critical paths and with a low logic depth are most sensitive to uncorrelated gate delay variations [15, 16].

As a first approximation, the gate delay of an inverter can be described by

$$T_{\text{gate}} \propto \frac{C_{\text{load}} V_{\text{DD}}}{\mu C_{\text{OX}} (W/L)(V_{\text{DD}} - V_{\text{th}})^{\alpha}}, \tag{3.1}$$

where μ is an effective mobility, V_{th} is the threshold voltage, C_{ox} is the gate capacitance per unit area, W and L are the transistor dimensions, and α is a power approximation coefficient that has a value between 2 and 1, with respect to the short channel effect [17]. A reduced supply voltage (V_{DD}) leads to an increased sensitivity $S_{T_{\text{gate}}}^{V_{\text{th}}}$ of gate delays to parameter variations [13]:

$$S_{T_{\text{gate}}}^{V_{\text{th}}} = \frac{V_{\text{th}}}{T_{\text{gate}}} \cdot \frac{\partial T_{\text{gate}}}{\partial V_{\text{th}}} = \frac{\alpha V_{\text{th}}}{V_{\text{DD}} - V_{\text{th}}}. \tag{3.2}$$

Furthermore, small transistor dimensions increase the effect of geometry-dependent parameter variations (variations in effective channel lengths). The impact of local variations may become significant since small-dimension devices operating at low supply voltages show an increased sensitivity to parameter variations. Therefore, for the design of low-voltage digital circuits the effect of intradie local parameter variations has to be minimized.

In this section, we present a novel technique that minimizes the impact of WID variations using redundancy applied only on critical parts of the circuit. First, the impact of WID parameter variations on gate and path delays is discussed with respect to critical path length. Moreover, a maximal critical path delay distribution has been derived. In the next section local delay variation minimization techniques, including majority and averaging gate, are presented. Last, a global analysis on the chip level is given followed by some conclusions.

Within-Die Parameter Variations and Their Effect on Gate and Path Delays

The two main parameters influencing gate delays are V_{th} and variations in effective channel length [18]. The impact of scaling on supply voltage and V_{th} mismatch variations is shown in Fig. 3.10. The data are obtained from [19]. With recent technology

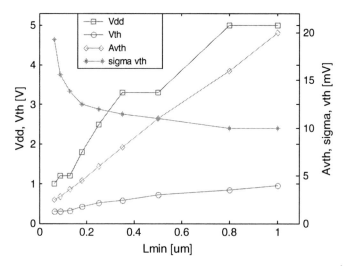

Fig. 3.10 Supply voltage (V_{DD}), threshold voltage (V_{th}), V_{th} mismatch coefficient $(A_{V_{\text{th}}})$, and standard deviation of V_{th} mismatch $(\sigma_{V_{\text{th}}})$ in different technology nodes

nodes (90, 65, and 45 nm) V_{th} mismatch ($\sigma_{V_{th}}$) value is significantly increased. The V_{th} impact of WID variations becomes very prominent with further reduction in the supply voltage.

Impact of the Critical Path Length and Gate Correlation on Delay Distribution

The critical path delay distributions resulting from D2D and WID parameter variations are calculated from D2D and WID statistical models using a SPICE-equivalent circuit simulator [20] with a 65-nm process file and a netlist containing modeled critical paths from a microprocessor.

To model a critical logic path in a design, a simple static inverter chain is used, containing a number n_{cp} of identical inverters, where each inverter drives four copies of itself yielding an FO4 load. FO4 is a common metric for the first-order analysis and evaluation of digital circuit performance in given process technologies [21]. The critical path delay $\left(T_{cp}\right)$ is calculated as

$$T_{cp} = n_{cp} T_{inv}, \qquad (3.3)$$

where T_{inv} is an average propagation delay through an FO4 inverter.

The WID variations model represents systematic WID parameter variations by expressing the device-to-device correlation as a function of the distance between the devices. This correlation function, however, is significantly influenced by specific manufacturing capabilities. For future technology nodes relatively smaller gate-to-gate correlation factors are expected, considering that wire connections do not scale with the same factor as transistor sizes, which makes gate-to-gate transistor distances relatively larger with respect to previous technologies.

To simplify the analysis, two separate WID variations cases can be identified: (1) completely dependent gates (gate delay correlation equal to 1 and (2) completely independent gates (gate delay correlation equal to 0), which may be viewed as extreme conditions of systematic and random variations, respectively.

In the completely systematic case, the variations have the same impact on every element in a critical path:

$$\frac{\sigma_{T_{cp}}}{T_{cp}} = \frac{n_{cp} \sigma_{T_{inv}}}{n_{cp} T_{inv}} = \frac{\sigma_{T_{inv}}}{T_{inv}}, \qquad (3.4)$$

where $\sigma_{T_{cp}}$ and $\sigma_{T_{inv}}$ are the standard deviations of the critical path delay distribution and the inverter gate delay distribution, respectively. This case, however, is not realistic in state-of-the-art technology nodes, where the distance between transistor gates relative to transistor sizes increases with respect to scaling.

In the case of completely random variations, however, the variations in the critical path delay are expected to have an averaging effect over the gates in the path [15]:

$$\frac{\sigma_{T_{cp}}}{T_{cp}} = \frac{\sigma_{T_{inv}}}{\sqrt{n_{cp}}T_{inv}}. \tag{3.5}$$

For completely random WID variations, the ratio of standard deviation to mean for the critical path delay distribution is inversely proportional to the square root of n_{cp} [13,15]. This is a more realistic approximation of the circuit delay variations in state-of-the-art technology nodes than (3.4). To demonstrate this, the general case is compared to the simulated data. In the general case, the variance of the sum of n_{cp} identical random variables $(X_1, \ldots, X_n)$ is given by [22]

$$\mathrm{Var}\left(\sum_{i=1}^{n_{cp}} X_i\right) = \sum_{i=1}^{n_{cp}} \mathrm{Var}(X_i) + 2\sum_{i=1}^{n_{cp}-1}\sum_{j=i+1}^{n_{cp}} \sqrt{\mathrm{Var}(X_i)\mathrm{Var}(X_j)}\rho(X_i,X_j), \tag{3.6}$$

where $\rho(X_i, X_j)$ is a correlation factor between random variables X_i and X_j. If (3.6) is applied to gates in the critical path, $\mathrm{Var}(X_i)$ becomes $\sigma^2_{T_{inv}}$ for every gate, $\mathrm{Var}\left(\sum_{i=1}^{n_{cp}} X_i\right)$ becomes $\sigma^2_{T_{cp}}$, and $\rho(X_i, X_j)$ becomes $\rho_{i,j}$ – the correlation factor between the ith and jth gates in the critical path. With these substitutions (3.6) becomes

$$\sigma^2_{T_{cp}} = n_{cp}\sigma^2_{T_{inv}} + \sigma^2_{T_{inv}} \times \sum_{i=1}^{n_{cp}-1}\sum_{j=i+1}^{n_{cp}} \rho_{i.j}. \tag{3.7}$$

After dividing (3.7) by T_{cp}, with respect to (3.3),

$$\frac{\sigma_{T_{cp}}}{T_{cp}} = \frac{\sigma_{T_{inv}}}{\sqrt{n_{cp}}T_{inv}}\sqrt{1 + \frac{2}{n_{cp}}\sum_{i=1}^{n_{cp}-1}\sum_{j=i+1}^{n_{cp}} \rho_{i.j}}. \tag{3.8}$$

If $\rho_{i,j} = 1$ for any pair of gates, then (3.7) becomes (3.4), otherwise for $\rho_{i,j} = 0$ for any pair of gates, (3.7) becomes (3.5). Due to the difference in their physical origins, variations of L and V_{th} exhibit different characteristics of correlation among transistors: lithography-induced variation of L is spatially correlated [23], while the V_{th} variation is mostly random due to dopant fluctuations [24]. For a small-size gate, a strong correlation is usually assumed to reduce the complexity of analysis. However, for a realistic circuit path that spans across larger distances, knowing the spatial correlation among gates is important for accurate statistical timing analysis [25]. A spatial correlation can be modeled as a linear function of distance [23]. Here we are using the following simplified model:

$$\rho_{i,j} = \begin{cases} \rho_0\left(1 - \frac{j-i-1}{D}\right) & \text{for } j-i \leq D \\ 0 & \text{for } j-i > D \end{cases}, \tag{3.9}$$

Fig. 3.11 Spatial correlation modeled as a linear function of distance

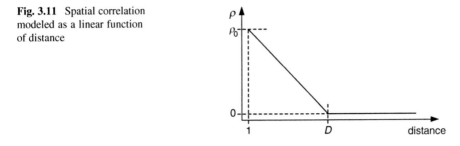

Table 3.1 Ratio of standard deviation to mean delay for WID and D2D variations and different critical path lengths (n_{cp})

$\sigma_{T_{cp}}/\mu_{T_{cp}}$ (%)	$n_{cp} = 1$	$n_{cp} = 6$	$n_{cp} = 8$	$n_{cp} = 10$
WID	11.48	5.44	4.78	4.44
D2D	7.96	7.94	8.12	8.22
	$n_{cp} = 12$	$n_{cp} = 16$	$n_{cp} = 20$	$n_{cp} = 24$
WID	4.08	3.62	3.32	2.98
D2D	8.29	8.38	8.43	8.46

where ρ_0 is the correlation factor between neighboring gates ($j = i + 1$) and D is the maximum distance between gates where correlation effects are still present. A spatial correlation model from (3.9) is illustrated in Fig. 3.11.

WID and D2D variations are acquired from Monte Carlo SPICE simulations using mismatch and parameter models for commercial 65-nm technology. Table 3.1 summarizes the statistical simulations for different critical path lengths providing the ratio of the standard deviation to the mean delay $(\sigma_{T_{cp}}/\mu_{T_{cp}})$ corresponding to WID and D2D variations.

An almost constant value for the standard deviation of D2D variations for different critical path lengths confirms that D2D variations can be assumed as purely systematic.

Substituting (3.9) into (3.8) with assumption that $n_{cp} \leq D + 2$ yields

$$\rho_0 = \frac{\left(\frac{\sigma_{T_{cp}} T_{inv}}{T_{cp}\sigma_{T_{inv}}}\right)^2 n_{cp} - 1}{(n_{cp} - 1)\left(1 - \frac{n_{cp}-2}{3D}\right)}. \qquad (3.10)$$

Eqation (3.10) has two parameters, ρ_0 and D, that need to be estimated for the given technology. Linear fitting is applied to (3.10) using the values for WID variations from Table 3.1. The following values are acquired (95% confidence parameter interval in brackets): $\rho_0 = 0.08$ (0.078–0.082) and $D = 8$ (7.3–8.3). Such a low correlation factor even for neighboring gates justifies the assumption that WID variations are mostly random.

The mean delay has been taken as a nominal critical path delay $T_{cp,nom}$. The WID and D2D nominal critical path standard deviations are $\sigma_{T_{cp},WID}$ and $\sigma_{T_{cp},D2D}$, respectively. The critical path delay probability density functions (PDFs) resulting from WID and D2D parameter variations are modeled as normal distributions [15, 18, 26] in (3.11) and (3.12), respectively:

$$f_{T_{cp,nom},WID} = N(T_{cp,nom}, \sigma^2_{T_{cp},WID}),\qquad(3.11)$$

$$f_{T_{cp,nom},D2D} = N(T_{cp,nom}, \sigma^2_{T_{cp},D2D}).\qquad(3.12)$$

Impact of Within-Die Variations on the Maximum Critical Path Delay Distribution

Following (3.11) and the procedure from [15] the probability of one critical path satisfying a specified maximum delay T_{max} is given with the cumulative distribution function (CDF) in (3.13):

$$F_{T_{cp,nom},WID}(T_{max}) = P_{T_{cp,nom},WID}(t < T_{max}) = \int_0^{T_{max}} f_{T_{cp,nom},WID}dt.\qquad(3.13)$$

A large chip, however, contains *many* critical paths, all of which must satisfy the worst-case delay constraint [13–16,27]. For completely dependent paths (path delay correlation equal to 1), the PDF given in (3.11) is valid for all the paths to model the worst-case delay. However, thanks to a very low gate correlation factor, as shown in the previous section, all the paths can be assumed to be independent (path delay correlation equal to 0). Assuming a number N_{cp} of independent critical paths for the entire chip [13], the probability that the whole chip satisfies the worst-case delay is given with a CDF for the whole chip $(F_{chip,WID})$ in (3.14):

$$F_{chip,WID}(T_{max}) = P_{chip,WID}(t < T_{max}) = \left(F_{T_{cp,nom},WID}(T_{max})\right)^{N_{cp}}.\qquad(3.14)$$

The chip's WID maximum critical path delay PDF is then calculated following [15] by taking the derivative of CDF with respect to T_{max}:

$$f_{chip,WID}(T_{max}) = \frac{dF_{chip,WID}(T_{max})}{dT_{max}}$$
$$= N_{cp} f_{T_{cp,nom},WID}(T_{max}) \left(F_{T_{cp,nom},WID}(T_{max})\right)^{N_{cp}-1}.\qquad(3.15)$$

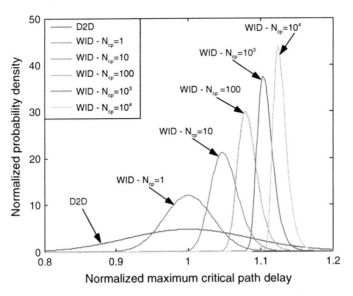

Fig. 3.12 Within-die (WID) maximum critical path delay distribution for different values of N_{cp} and die-to-die (D2D) critical path delay distribution

The chip's maximum critical path delay PDF (3.15) is illustrated in Fig. 3.12
   By increasing N_{cp},

- The mean of $f_{chip,WID}$ increases, since the *slowest* critical path limits the chip's overall performance and the probability of a longer cycle time increases;
- The standard deviation of $f_{chip,WID}$ decreases and becomes relatively small compared to the standard deviation of $f_{T_{cp},nom,WID}$, e.g., for $N_{cp} = 10^4$, $\sigma_{chip,WID} = 0.3\sigma_{T_{cp},nom,WID}$, and $\sigma_{chip,WID} = 0.12\sigma_{T_{cp},nom,D2D}$;
- $f_{chip,WID}$ becomes less sensitive to further increases of N_{cp}; qualitatively, this means that an increase of N_{cp} from 1 to 10 has a greater effect on the mean and variance of the WID distribution than an increase from 10^3 to 10^4;
- The shape of $f_{chip,WID}$ becomes less symmetrical and more positively skewed.

As the number of transistors per chip increases and the number of average gate delays per critical path is reduced [28], N_{cp} is expected to increase for each technology generation and further reduce the sensitivity of maximum critical delay to N_{cp}.

Local Delay Variation Minimization Techniques

Redundancy can be used to reduce the delay variance of critical paths and thereby increase the overall circuit speed. The general principle consists in replicating R

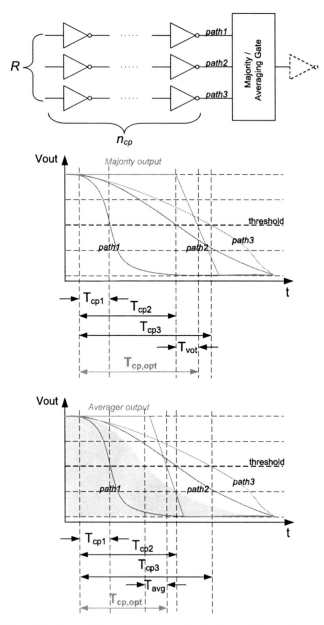

Fig. 3.13 Schematic of circuit for local delay minimization and signal timing diagrams for majority and averager gate realizations

times a critical path and evaluating critical path outputs using a function that delivers the output with reduced time delay variance. Functions that satisfy this condition, and that are considered in this paper, are realized using majority $(R + 1)/2$ out of R and averaging the voters. The principle of operation is illustrated in Fig. 3.13. For a majority gate, a voter circuit switches when a path with a median value of time delay reaches a selected threshold voltage. For an averaging gate, a circuit switches when the average value of the outputs of all the paths reaches a selected threshold voltage.

The proposed technique can also be used to support recovering the correct operation, which is disrupted by other sources of variations and signal-integrity issues such as those caused by crosstalk aggressor signals.

Majority Gate Delay Variation Minimization

A majority gate performs a median function of its inputs. The median of a statistical distribution with CDF $D(x)$ is the value of x such that $D(x) = 1/2$. For a symmetric distribution, it is therefore equal to the mean. Having the order statistics $Y_1 = \min_j X_j$, $Y_2, \ldots, Y_{R-1}$, $Y_R = \max_j X_j$ gives the statistical median of the random sample (3.16) [22]:

$$\tilde{x} \equiv \begin{cases} Y_{(R+1)/2} & \text{if } R \text{ is odd} \\ \frac{1}{x}\left(Y_{R/2} + Y_{1+R/2}\right) & \text{if } R \text{ is even} \end{cases} \tag{3.16}$$

Only odd values of R are used since the majority gate can only support an *odd* number of inputs.

Taking into consideration that random variables $X_1, \ldots, X_R$ follow a normal distribution with mean μ and standard deviation σ, $\tilde{x}$ also has a normal distribution with the mean (μ_{med}) equal to μ and standard deviation (σ_{med}) that has an asymptotic upper bound (for $R \to \infty$) equal to $\sqrt{\frac{\pi}{2}}\frac{\sigma}{\sqrt{R}}$ [29]. There are various estimations for the standard deviation [29–31]; nevertheless, we consider that a much simpler formula given in (3.17) provides the most appropriate estimation for all cases where a low value of R ($R < 10$) is considered. In Table 3.2 the exact values obtained from [29], approximation values from [30, 31], and the values obtained using our approximation (3.17) are given for $\sigma = 1$:

Table 3.2 Different estimations of standard deviation of median function for various redundancy factors (R)

	Sample size				
	$R = 3$	$R = 5$	$R = 7$	$R = 9$	$R = 11$
Exact SD	0.6692	0.5356	0.4587	0.4075	0.3704
Our app. of SD	0.6699	0.5344	0.4576	0.4066	0.3696
SD from [30]	0.6202	0.5134	0.4466	0.4001	0.3654
SD from [31]	0.6637	0.5337	0.4580	0.4072	0.3701

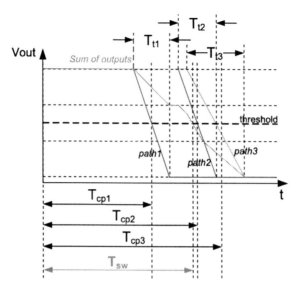

Fig. 3.14 Linearized signals for critical path delays and a switching point delay

$$\sigma_{\mathrm{med}} = \sqrt{\frac{\pi}{2R+1}}\sigma. \tag{3.17}$$

We note that in a real implementation, the additional majority block may also add to the overall delay variation; however, this will not be taken into consideration.

Minimization of Averaging Gate Delay Variations

The averaging gate switches when the arithmetic average of its input levels reaches the threshold ($V_{\mathrm{DD}}/2$). The switching point depends on the path propagation delay $(T_{\mathrm{cp},i})$ and the last gate transition delay $(T_{t,i})$ for each path ($i = 1, \ldots, R$). The expression for the switching point delay for linearized signals (Fig. 3.14) is given in (3.18), where m is the number of inputs whose transitions are completed before the switching point and n is the number of inputs whose transitions start after the switching point, as illustrated in Fig. 3.14, where $m = 1, n = 0$, and $R = 3$:

$$T_{\mathrm{sw}} = \frac{\frac{(m-n)}{2} + \sum\limits_{i=n+1}^{R-m} \frac{T_{\mathrm{cp},i}}{T_{t,i}}}{\sum\limits_{i=n+1}^{R-m} \frac{1}{T_{t,i}}}. \tag{3.18}$$

Taking into consideration that $T_{\mathrm{cp},i}$ and $T_{t,i}$ for every $i = 1, \ldots, R$ are random variables following a normal distribution with mean $\mu_{T_{\mathrm{cp}}}$ and μ_{T_t} and standard deviation $\sigma_{T_{\mathrm{cp}}}$ and σ_{T_t} respectively, T_{sw} also has a normal distribution with a mean (μ_{sw}) that

Table 3.3 Ratio of standard deviation over mean delay for different redundancy factors (R) and critical path lengths (n_{cp}) for majority and averaging gate realizations

%	$\frac{\sigma_{Tcp}}{\mu_{Tcp}}$	$R = 3$		$R = 5$		$R = 7$	
		$\frac{\sigma_{med}}{\mu_{med}}$	$\frac{\sigma_{sw}}{\mu_{sw}}$	$\frac{\sigma_{med}}{\mu_{med}}$	$\frac{\sigma_{sw}}{\mu_{sw}}$	$\frac{\sigma_{med}}{\mu_{med}}$	$\frac{\sigma_{sw}}{\mu_{sw}}$
$n_{cp} = 6$	5.44	3.64	3.17	2.91	2.46	2.49	2.09
$n_{cp} = 8$	4.78	3.20	2.81	2.56	2.19	2.19	1.85
$n_{cp} = 10$	4.44	2.97	2.63	2.37	2.05	2.03	1.73
$n_{cp} = 12$	4.08	2.73	2.44	2.18	1.89	1.87	1.60
$n_{cp} = 16$	3.62	2.43	2.21	1.93	1.71	1.66	1.45
$n_{cp} = 20$	3.32	2.22	2.05	1.78	1.59	1.52	1.34
$n_{cp} = 24$	2.98	1.99	1.85	1.59	1.44	1.36	1.22

is approximated with μ_{Tcp} and a standard deviation (σ_{sw}) that has an upper bound equal to

$$\sigma_{sw} = \frac{\sigma_{Tcp}}{\sqrt{R}}. \tag{3.19}$$

The exact value of the mean and standard deviations for a normal distribution of T_{sw} cannot be derived analytically. Therefore, values representing the ratio of the standard deviation to the mean delay obtained by Monte Carlo simulations are shown in Table 3.3 for different critical path lengths (n_{cp}) and different redundancy factors (R). The values related to the majority gate are also acquired from Monte Carlo simulations for the sake of comparison, and we observe that they comply with the approximation given in (3.17). The values related to the averaging gate are always close to the upper bound (3.19), thereby demonstrating that the averaging gate performs an optimal minimization of the standard deviation of critical paths delay. The mean and standard deviation values for a critical path and a transition time for the last gate in the path, μ_{Tcp}, μ_{T_i}, σ_{Tcp}, and σ_{T_i} respectively, are given for a 65-nm fabrication technology.

The averaging function has two prominent advantages over the majority function, namely, it enables better minimization of the standard deviation of the output delay; moreover, the averaging function has no restriction on redundancy factor R, whereas the majority function demands odd values only.

Optimization Method Using the Proposed Techniques for Variation Minimization

The effect of reducing the standard deviation of critical paths by replicating a critical path and inserting a decision gate is exploitable only when the reduction in the standard deviation is large enough to compensate for the additional delay of the added decision gate. As previously shown, longer critical path lengths are beneficial for the minimization of intradie delay variations thanks to the averaging effect over the gates in the critical path.

An architecture that has been optimized in terms of delay variations is considered in what follows as a multithreaded processor with $N_{cp} = 10^4$ critical paths. N_{cp} is estimated by assuming that the ratio between the number of independent critical paths and the number of transistors per chip remains approximately constant in different technologies [15]. The exact number is not significant since the sensitivity of the maximum critical path distribution to N_{cp} is small for large values of N_{cp} $\left(N_{cp} \sim 10^3\text{–}10^4\right)$, as demonstrated. As shown in [32], an optimal critical path length for power/performance in a multithreaded processor is in a range of 20 to 24 equivalent FO4 inverter delays. Therefore, cases where $n_{cp} = 20$ and 24 are taken into consideration in what follows. The distribution for a normalized maximum critical path delay is shown in Fig. 3.15, considering cases with and without optimization. The optimization is performed with *averaging* gates and a redundancy factor $R = 3$.

Figure 3.15 clearly shows that longer critical paths have a reduced mean value of maximum critical path delay and are therefore less sensitive to WID variations. A reduced standard deviation also reduces the mean value of the maximum critical path delay. However, adding a majority/averaging gate increases it. For the given technology, any path where $n_{cp} \geq 24$ has a smaller or equal normalized mean value of the maximum critical path delay when the proposed optimization technique is applied, compared to a default case (a case without any delay variations minimization technique) $\left(\mu_{T_{cp.max.opt}} / T_{cp.nom} = \mu_{T_{cp.max}} / T_{cp.nom} = 1.115\right)$. However, the standard deviation is smaller $\left(\sigma_{T_{cp.max.opt}} / T_{cp.nom} = 0.58\% < \sigma_{T_{cp.max}} / T_{cp.nom} = 0.91\%\right)$, which can give better yield. This is investigated in detail in the section "Adaptive

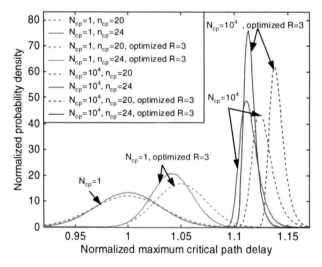

Fig. 3.15 Within-die (WID) maximum critical path delay distribution for $n_{cp} = 20$ and $n_{cp} = 24$ without and with optimization ($R = 3$)

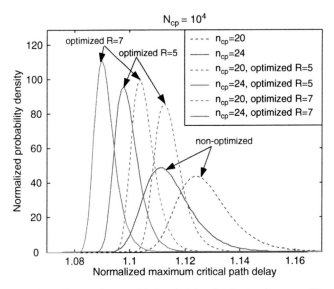

Fig. 3.16 Within-die (WID) maximum critical path delay distribution for $n_{cp} = 20$ and $n_{cp} = 24$ without and with optimization $\left(N_{cp} = 10,000, R = 5\right)$

V_{gs}: A Novel Technique for Controlling the Power and Delay of Logic Gates in a Sub-VT Regime," where D2D variations are also considered.

When $n_{cp} = 20$, the normalized mean value of the maximum critical path delay is larger if the proposed optimization technique is applied ($\mu_{T_{cp.max.opt}} / T_{cp.nom} = 1.131 > \mu_{T_{cp.max}} / T_{cp.nom} = 1.128$). In order to achieve a noticeable improvement in optimization performance, the redundancy factor of critical paths needs to be increased. The distribution of a normalized maximum critical path delay is shown in Fig. 3.16, which considers the cases where optimization uses redundancy factors of 5 and 7. The improvement in the mean value of maximum critical path delay is

- 1.5 and 1.35% compared to the default case (without optimization) for $n_{cp} = 24$ and $n_{cp} = 20$, respectively, when a redundancy factor of 5 is chosen;
- 2.3 and 2.2% for $n_{cp} = 24$ and $n_{cp} = 20$, respectively, when a redundancy factor of 7 is chosen.

Maximum Critical Path Delay Distribution with Combined Die-to-Die and Within-Die Variations

The maximum critical path delay distribution of the chip considering both D2D and WID types of variations can be obtained by combining the individual D2D and WID distributions, following an adapted version of the procedure presented in [15]. The maximum critical path delay is calculated as

$$T_{cp,max} = T_{cp,nom} + \Delta T_{cp,WID} + \Delta T_{cp,D2D} = T_{cp,WID} + \Delta T_{cp,D2D}, \qquad (3.20)$$

where $\Delta T_{cp,D2D}$ and $\Delta T_{cp,WID}$ are the deviations in the nominal critical path delay resulting from D2D and WID variations, respectively. The maximum critical path delay density function resulting from both D2D and WID variations is derived using convolution according to (3.20):

$$f_{T_{cp,max}} = f_{T_{cp,nom},WID} * f_{\Delta T_{cp,max},D2D}, \qquad (3.21)$$

where $f_{T_{cp,nom},WID}$ is as given in (3.10) and $f_{\Delta T_{cp,max},D2D}$ is a distribution resulting from shifting in the negative direction the D2D distribution (3.11) by $T_{cp,nom}$ expressed as

$$f_{T_{cp,nom},D2D} = N(0, \sigma^2_{T_{cp,D2D}}). \qquad (3.22)$$

Since the WID distribution has a significantly smaller standard deviation compared to the D2D distribution, which is further reduced when N_{cp} increases, the WID distribution can be legitimately approximated with an impulse function. As the D2D and WID distributions are statistically combined through (3.21), the resulting distribution has a mean equal to that of the WID distribution and a variance predominantly resulting from the D2D distribution. Thus, WID variations determine the mean of the maximum critical path delay distribution, and D2D variations determine the variance. With respect to this observation, any improvement in the mean value of the maximum critical path delay WID variations causes a direct improvement in the maximum critical path delay of any fabricated chip. A tradeoff can be considered between the redundancy level involved, which represents additional area/power, and a reduction of the maximum critical path delay, which actually means possible increased operating frequency. Assuming that each gate in the critical path consists of six transistors, $N_{cp} = 10^4$, and that a whole chip has 4×10^8 transistors according to ITRS [10], replicating each critical path 3, 5, and 7 times causes an overhead of 0.8, 1.5, and 2.3% respectively.

The maximum critical path delay distribution, including both WID and D2D variations, is analyzed for three technologies: one 65-nm commercial technology and two future technology nodes 45 and 32 nm. Both the 45- and 32-nm technologies are hypothetical, and the parameters have been selected to anticipate a realistic set [10, 18, 33–36].

Values for the ratio of the standard deviation to the mean delay (σ/μ) for WID, D2D variations, and for critical path length (n_{cp}) are given in Table 3.4. Standard deviation values are estimated by combining $\sigma_{V_{th}}$ and σ_L according to [18, 33–36], and n_{cp} is estimated as presented in [32]. $N_{cp} = 10^4$; finally, $R = 5$ in all calculations.

The maximum critical path delay distribution of the chip for each technology node is depicted in Fig. 3.17 and is calculated using the values from Table 3.4.

The improvement of the mean value of maximum critical path delay is equal to 1.5, 3.2, and 4.3% for the 65-, 45-, and 32-nm technologies, respectively. The larger improvement than overhead due to redundancy for future technologies suggests

Table 3.4 Values of σ/μ for gate and critical path WID and D2D variations and n_{cp} for different technologies

Technology (nm)	D2D (%): σ/μ	WID (%): σ_{gate}/μ_{gate}	WID (%): $\sigma_{T_{cp}}/\mu_{T_{cp}}$	n_{cp}
65	8.4	11.5	3	24
45	10	15	3.5	30
32	12	20	4	36

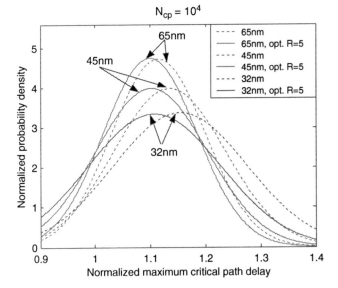

Fig. 3.17 Maximum critical path delay distribution for combined D2D and WID variations for different technologies ($N_{cp} = 10,000$, $R = 5$)

that there will be an optimal design point where this technique for delay variation minimization can be successfully applied.

Summary

This section studies the benefit of using the averaging technique to limit the impact of delay variations. Another popular fault-tolerant technique (majority voting) is used for the sake of comparison. The averaging technique has two prominent advantages over the majority technique, namely, it enables better minimization of the standard deviation of the output delay and, moreover, when using the averaging mechanism, the redundancy factor can be as low as 2. The technique is intended to reduce the effects of intradie variations using redundancy applied only on critical segments (critical paths). This way, the proposed technique can be optimally used in the design of large synchronous digital systems.

The studies have shown that the technique can be already applied for a 65-nm CMOS technology process. However, the real benefit is expected for future nano-scale CMOS technologies such as 45- and 32-nm nodes where an optimal point has been shown to exist in the speed vs. area/power tradeoff.

Adaptive V_{gs}: A Novel Technique for Controlling the Power and Delay of Logic Gates in a Sub-VT Regime

The primary motivation for ultra-low-voltage operation is to reduce energy [37]. Analysis in [38] and chip measurements in [39] showed that minimum energy per operation occurs in the sub-VT region. An 8-T sub-VT SRAM in 65-nm CMOS is demonstrated in [40], and more complex sub-VT processors are appearing [41]. In a sub-VT region, with further supply voltage (VDD) scaling, gate delay and clock period increase exponentially, dynamic energy per operation decreases in a quadratic manner, but leakage power accumulates over the longer clock period, and finally leakage energy per operation exceeds the dynamic energy and causes the minimum energy point.

References [39–41] have used static CMOS gates. These gates continue to function in the sub-VT region and have a great potential for saving energy, but they face many challenges including temperature sensitivity, PV [42], and process imbalance [43].

There are four main sources of leakage current in digital circuits, i.e., reverse diode current (I_{diode}), sub-VT current (I_{SUB}), gate leakage (I_G), and GIDL current (I_{GIDL}). Usually I_{diode} is negligible. However, since the I_G and I_{GIDL} currents are exponential functions of supply voltage while I_{SUB} is a weak function of VDD (through DIBL effect), I_{SUB} is the dominant leakage term for the weak-inversion and sub-VT regions. There is no theoretical solution for I_{SUB} leakage. A commonly used expression for sub-VT current is given by [44]:

$$I_{sub} = \mu_0 C_{ox} \frac{W}{L} (n-1) V_{th}^2 \times e^{\frac{V_{gs} - V_{T0} - \gamma V_{sb} + \eta V_{ds}}{n V_{th}}} \times (1 - e^{\frac{-V_{ds}}{V_{th}}}), \quad (3.23)$$

where V_{th} is the thermal voltage (kT/q), n the sub-VT slop factor, η DIBL coefficient, and γ the body effect coefficient.

In this work, simulations are done using the transistor model card provided by the foundry for a 65-nm low-power process. Our simulations in 65-nm and chip measurement in [45] in 180 nm shows that, approximately independent of technology, leakage current (and speed) increases 2 × per 15°C temperature increase. One way to alleviate this strong temperature dependence is to change the frequency of operation as a function of temperature [45]. But this is not acceptable for most digital applications. Dynamic power is not a function of temperature. So I_{SUB} is the main source of power consumption at high temperatures.

Because of the exponential dependency of transistor current on the parameter variations, it is clear that any logic style designed for sub-VT operation should work sizing independently. Contention current between pull-up and pull-down networks (PUN and PDN) fails to work in the presence of intra-die variations.

Available Variation Compensating Techniques

In the sub-VT region, all kinds of parameter variations like TOX, channel length (L), V_T, and temperature variations show similar behavior. When they increase the I_{SUB}, they decrease the delay, and vice versa. So all of the circuit techniques that can change the power-delay tradeoff at runtime can be used for compensating all of these variations at the same time and independently of the variation source.

For over-100-nm technologies, adaptive body biasing (ABB) is a good technique for compensating the variations [46, 47]. Both forward body biasing (FBB) and reverse body biasing (RBB) can be used in a sub-VT regime. Since ABB changes the V_T value directly, it can control both leakage and delay. Also, the overhead of this technique is small. This technique is very good but has three important weaknesses. First, using ABB for compensating intradie variations of NMOS transistors need triple-well technology. Second, the increased SCE due to scaling decreases the body factor of bulk-CMOS drastically. According to the foundry data, with 65-nm technology, RBB can change the V_T value effectively to less than 60 mV. This amount causes an approx. fourfold change in delay and power in the sub-VT region, which is much less than PV and temperature effects. Also, much more than a 60-mV change in V_T is required for compensating slow and fast corners. And third, as we will discuss in the section "Minimizing Local Delay Variations for Nanoscale CMOS Technologies," the body factor is almost 0 in emerging multigate devices, which are promising candidates for future electronics.

Both power and gate delay strongly depend on VDD. So adjusting the VDD can be another technique for compensating variations. This can be done by using a variable supply voltage (Var-VDD). But this method does not provide a good control on I_{SUB} because this leakage current is a weak function of VDD through a second-order effect, i.e., DIBL.

Body Effect in Emerging Multigate Devices

For sub-50-nm technology, various nonplanar device structures have been explored for better SCE immunity and gate electrostatic control of the channel surface potential. Among the many approaches, double-gated FinFET, trigated, Π-gated, Ω-gated, NW body, and GAA MOSFETs have attracted much attention.

The GAA structure in which the gate oxide and the gate electrodes wrap around the channel region exhibit excellent electrostatic control, e.g., near ideal sub-VT slope (S) (<63 mV/dec), very low DIBL (<10 mV/V), and with an I_{ON}/I_{OFF} ratio of $<10^6$ [48].

The body effect coefficient or body factor, γ, represents the dependency of the V_T on the back-gate bias. For any devices, one can write $\gamma = -d(V_T)/d(V_b)$, where V_b is the back-gate voltage. The body factor can be calculated using a simple capacitive equivalent circuit and the relationship $\gamma \propto C_{CH-B}/C_{G-CH}$, where C_{CH-B} is the capacitance between the channel surface and back-gate (or the substrate) and is the capacitance between the gate electrode and the channel [49].

In [49] and [50] it is shown that by increasing the gate electrode control in these devices from single-gate $\rightarrow$ double-gate $\rightarrow$ trigate $\rightarrow$ Π-gate $\rightarrow$ GAA, DIBL, V_T roll-off by channel-length, and S-factor decrease. All of these reductions are considered as good effect for logic circuits and cause better performance and less power consumption, but unfortunately γ also decreases in the same direction.

S-factor is very important for sub-VT operation. For example, in 65-nm planar-CMOS S-factor is <93 mV/dec, but in the GAA devices of [48] and multigate FETs of [51] it is <63 mV/dec. So in the sub-VT region for exactly equal device performance: I_{ON}, I_{OFF}, and $I_{ON}/I_{OFF} = 10^n$, we have $VDD_{GAA}/VDD_{PLANAR} \approx (63$ mV $\times$ n$)/(93$ mV $\times$ n$)$. At the minimum energy point, the ratio of $Energy_{Leakage}/Energy_{Dynamic} \approx 1/3$. In logic, circuits for given I_{OFF} and I_{ON}, delay/$Energy_{Leakage}/Energy_{Dynamic}$, are linear/linear/quadratic functions of voltage swing (VDD), respectively. As a result, in the sub-VT regime using GAA devices instead of planar-CMOS causes approx. 1.9$\times$ less energy per operation and approx. 1.5$\times$ better performance (speed). This means that GAAs are the best choice for weak inversion and sub-VT operation.

It is interesting to note that from the circuit design point of view, the main problem of FinFETs is the low mobility in the saturation region because in the sidewalls the crystal orientation is (110) [51]. But for sub-VT operation μ_0 is not important because the current is dominated by diffusion. Variation of μ_0 in (1) can be compensated by V_{T0} adjustment.

In summary, in multigate devices, the body factor is much smaller than in single-gate devices because of the enhanced coupling between gate and channel and because the lateral gates shield the device from the electric field from the back gate [49]. Measurements in [48] show that in GAA devices the body factor is exactly 0.

In addition, the study in [52] and measurements in [53] show that the leakage current in all of these multigate devices is very PV and temperature sensitive. So we need to find new compensation techniques as replacements for ABB.

Proposed Adaptive V_{gs} Technique

Today's fabrication technologies offer the feasibility of using V_T-low, V_T-nominal, and V_T-high on the same die. The difference between V_T values ($\Delta V_T = V_T$-nominal - V_T-low $\approx V_T$-high - V_T-nominal) is usually 70 to 100 mV. So for sub-VT operations V_T-low devices are about ten times faster (and leakier) than V_T-nominal devices.

Among several available static logic styles, SCMOS and PTL are more popular. SCMOS is the simplest and most robust style and is especially good for designing simple functions like NAND and NOR. Designing PTL is more complicated but it has better performance than SCMOS in designing some complex functions like MUX and XOR. The best choice is using mixed PTL/SCMOS gates [54].

The AVGS technique requires four supply rails, two ground rails (ground and $0 + \Delta V_n$), and two power rails (VVDD and VVDD $- \Delta V_p$). VVDD and ground are the conventional rails. ΔV_n and ΔV_p are the new circuit parameters and on the order of a few tens of millivolts. In "Results and Discussions", we will propose a multioutput bulk-converter switching power supply that can generate all of these supply rails by using only one inductor.

AVGS–SCMOS

Figure 3.18a shows the proposed AVGS–SCMOS technique for a single-stage logic structure. All the input signals and the output signal F are full-swing ($0 \rightarrow$ VVDD). But the PDN and PUN are connected to $0 + \Delta V_n$ and VVDD $- \Delta V_p$, respectively. As a result, when the PDN is on, NMOS transistors have gate-source voltage $V_{gs} =$ VVDD $- \Delta V_n$, and when PDN is off $V_{gs} = -\Delta V_n$. Similarly for PUN, in the on state, the transistor drive voltage is $V_{sg} =$ VVDD $- \Delta V_p$, and in the off state $V_{sg} = -\Delta V_p$. In both on and off states, the transistor current changes exponentially according to (1). So $\Delta V_{n/p}$ can strongly control the leakage power and delay of the gate exactly the same as if we changed the transistor's threshold voltage.

But one needs a full-swing output signal for driving the next logic gate stages. The proposed 6-T level converting inverter (LCI) is also shown in Fig. 3.18a. This inverter generates the full-swing output signal F. When Do is high (VVDD $- \Delta V_p$), Mn0 and Mn1 are on, X_n and F are 0 V, and Mp2 is on, so $X_p =$ VVDD $- \Delta V_p$. Transistor Mp1 has $V_{gs} = 0$, so the leakage current through this transistor is the same as a conventional 2-T inverter at VDD $=$ VVDD $- \Delta V_p$. If $\Delta V_p > 0$, then Mp0 has $V_{sg} = V_{sd} = \Delta V_p$, so it is in diode connected mode; otherwise Mp0 is completely off. The voltage across this V_T-low diode is tens of millivolts, so the DIBL effect is negligible, but its leakage current can be high. Mp2 redirects this current to the VVDD $- \Delta V_p$ supply rail. This means that the voltage drop across this leakage path is very small, so its leakage power is small. If $\Delta V_n > 0$, then Mn2 is off, but if $\Delta V_n < 0$, then it works like a diode between ground and the $0 + \Delta V_n$ line, and because ΔV_n is small, its leakage power consumption will be small. Similarly when Do is low, $F =$ VVDD, $X_n = \Delta V_n$, and V_{gs} (Mn1) $= 0$. If $\Delta V_n > 0$, then V_{ds} (Mn0) $= V_{gs}$ (Mn0) $= \Delta V_n$ and it works like a diode. Mp2 and Mn2 do not have any other functionality. They are always minimum-size transistors and only provide paths for diode currents.

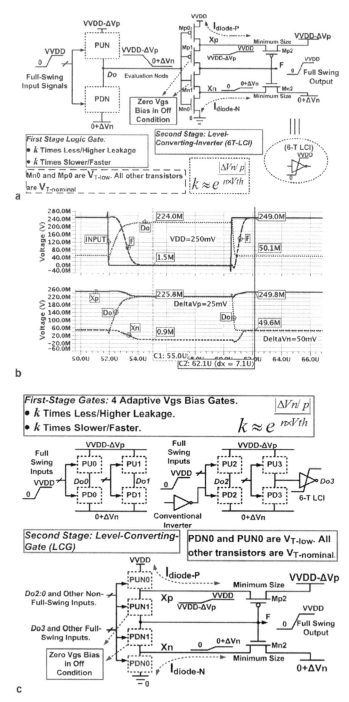

Fig. 3.18 Proposed AVGS style. **a** AVGS–SCMOS single-stage logic gate. **b** voltage waveforms of OR gate at VVDD = 250 mV, $\Delta V_{\text{n}} = 50$ mV, and $\Delta V_{\text{p}} = 25$ mV. **c** AVGS–SCMOS two-stage logic gate

Voltage waveforms of AVGS–SCMOS are shown in Fig. 3.18b. The voltage swings in X_n and X_p nodes are very small $(\Delta V_{n/p})$. So the effect of the parasitic capacitances of these nodes on the delay and dynamic power is very small. Mn0 and Mp0 are V_T-low devices and about ten times faster than Mn1 and Mp1, respectively. The W of these transistors can be smaller than Mn1 and Mp1. Because Mn0 and Mp0 contribute to only approx. 10% of 6-T LCI delay and their leakage power is small, the PV sensitivity of these devices is not important. Also, the effect of Mn2 and Mp2 on the PV sensitivity of timing and power is negligible because they only provide paths for leakage currents.

When we have several gate stages, we can decrease the overheads by using level converting gates (LCG) instead of LCIs. Figure 3.18c shows a general two-stage AVGS–SCMOS circuit. All of the first-stage gates are supplied by new power rails. LCG has exactly the same structure as LCI; PDN and PUN are duplicated using V_T-low devices. The LCG can have both full-swing and non-full-swing inputs. The voltage swing in X_n and X_p nodes is $\Delta V_{n/p}$. PDN0 and PUN0 are V_T-low, and PDN1 and PUN1 are V_T-nominal, so PDN0/PUN0 is much faster than PDN1/PUN1, respectively.

AVGS–PTL

Various PTL styles can be summarized in three general structures. The simplest way to implement a switch is using a transmission gate (TG). PTL implemented with TG is called PTL+ (also called CMOS+). This structure is very robust and provides full-swing input and output signaling and works sizing independent. Single-rail PTL and complementary PTL styles use NMOS switch networks and swing restoration is done by cross-coupled PMOS transistors (cross-coupled inverters in SRPL). Single-rail PTL and CPL are not suitable for the sub-VT region due to the sizing dependency and contention current between swing restorer and NMOS network.

One of the drawbacks of the PTL style is that some of the input signals are directly connected to source/drain junctions. This fact causes two problems. First, logic gate input capacitance is data dependent. In timing analysis we always have to consider worst-case delay, so this issue deteriorates the timing. Second, the charge sharing between interconnects and internal nodes as shown in Fig. 3.19a can increase the signal path delays several times. Also, it is difficult to model this effect in the CAD tools because slt1 and IN1 are two independent signals in two separate paths, path1 and path2. To eliminate both problems we can add inverters to buffer the input signals that drive the source/drains junctions. This solution is shown in Fig. 3.19b. Inputs that drive the transistor gates do not need to be buffered. As shown in Fig. 3.19b, it is also possible to have AVGS mixed SCMOS–PTL gates.

Figure 3.19c illustrates an example AVGS–PTL gate that calculates $F = (A + B) (C + D)$. As illustrated in Fig. 3.19d, it is quite easy to apply the AVGS technique to lookup tables (LUTs). Because charge sharing and crosstalk noise may change the SRAM internal data, we should add the buffers INV15:0. To the best of our

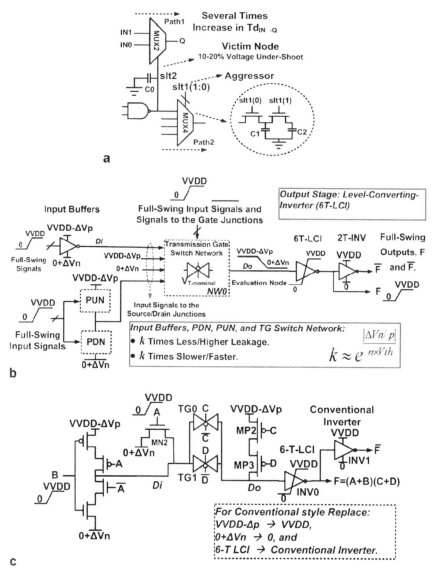

Fig. 3.19 Proposed AVGS style. **a** Charge-sharing and input-capacitance data-dependency problems of PTL. **b** Proposed AVGS–PTL logic style. **c** Function $F = (A + B)(C + D)$. **d** MUX-based sub-VT LUT4. **e** Voltage waveforms of LUT4 at VVDD = 300 mV, $\Delta V_n = 25$ mV, and $\Delta V_p = -25$ mV

knowledge, until now no work has been done on sub-VT FPGAs. Both FPGA and sub-VT circuits are considered as solutions for low-performance applications. Thus it is quite reasonable to have FPGAs with very low energy per operation. Example voltage waveforms are shown in Fig. 3.19e.

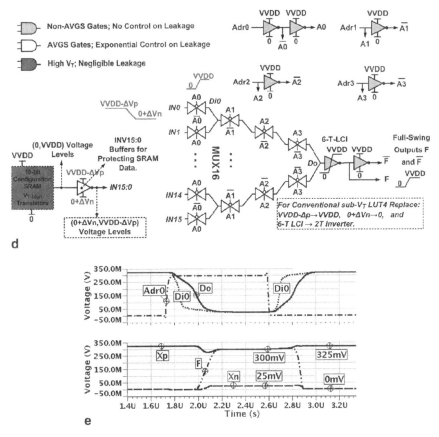

Fig. 3.19 Continued

Results and Discussions

Figure 3.20a–d shows the normalized power delay curves of AVGS–SCMOS, VAR-VDD, and ABB methods for an OR16 gate. SCMOS OR16 performance at VVDD= 300 mV is the reference. AVGS–SCMOS OR16 has four NOR4 in the first stage and level converter NAND4 in the second stage. In Fig. 3.20a, when $\Delta V_{n/p} = 0$, we see about 3% dynamic energy and 19% delay overheads because of ten extra transistors (25%). The dynamic energy overhead is small because switching activity in Do nodes is very small. By increasing $\Delta V_{(n/p)}$, delay increases and leakage decreases exponentially. It is not useful to increase $\Delta V_{n/p}$ more than +80 mV because the leakage curve will saturate to the leakage of LCG.

VAR-VDD and ABB are applied to conventional SCMOS OR16 gates. Figure 3.20b shows VAR-VDD behavior. Delay increases exponentially but power decreases in a quadratic form. As shown in Fig. 3.20c, RBB fails to save power at the TT corner. We have used an industrial 65-nm bulk-CMOS low-power process

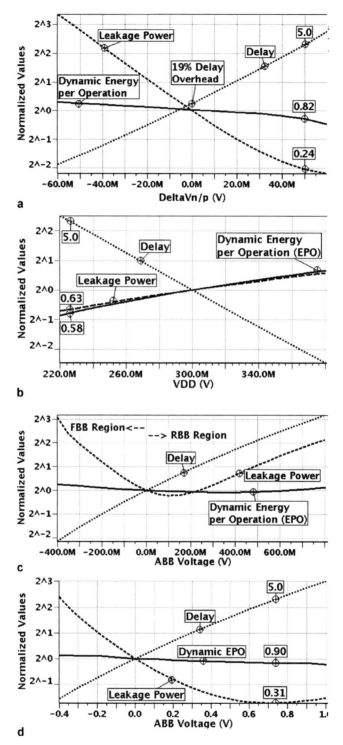

Fig. 3.20 Power-delay curves of OR16. **a** AVGS with $\Delta V_n = \Delta V_p$. **b** VAR-VDD. **c** ABB in TT corner. **d** ABB in FF corner

for simulations, and in this process I_{SUB} is comparable to other leakage terms. RBB increases V_T (and delay) but it also increases other leakage terms. But in the FF corner I_{SUB} is quite dominant and, as is shown in Fig. 3.20d, RBB can compensate this corner about 3.3×. AVGS and VAR-VDD work in all corners. The simulation results of LUT4 (Fig. 3.19d) are very similar to the power delay curves shown in Fig. 3.20a–d. So to save space, we omit them.

Figure 3.21 shows the ability of AVGS–PTL to compensate the process imbalance, FS, and SF corners. In this style, ΔV_n controls the NMOS transistors in NW0 and in input buffers (Fig. 3.19b), while ΔV_p controls the PMOS ones. So by increasing ΔV_n when ΔV_p is constant we can compensate the fast-NMOS slow-PMOS (FS) corner. Var-VDD cannot provide independent control on NMOS and PMOS transistors.

ABB and RVGS techniques can be combined together to provide more control over the power and delay. This is shown in Fig. 3.22. By using only an ABB/AVGS technique we can compensate the variations $3 \times /5.4\times$, respectively. But by using both techniques it is feasible to compensate 8.3× (RBB voltage $= 300\,\text{mV}$ for both NMOS and PMOS and $\Delta V_{n/p} = 50\,\text{mV}$). In this technology NMOS and PMOS transistors have approximately equal body-effect coefficients.

Figure 3.23 shows the total energy (drawn from all power supplies) per operation of the LUT4 structures shown in Fig. 3.19d. For the sake of simplicity, transistors Mn0 and Mp0 of 6T-LCI shown in Fig. 3.18a are implemented by V_T-nominal devices. The energy shown in Fig. 3.23 is the average value for a long random input data pattern. The switching activity (the probability of data transition in each clock cycle) of input address signal (Adr3:0) is 10%. The LUT4 critical path delay is from Adr0 to F and is 20% of the clock period. The FS corner causes a 71% increase at the minimum energy point and 5.5× speed variation in this point. Both AVGS and ABB techniques can compensate these variations about two times, and both need two extra power supplies.

This method provides exponential control on 60 to 70% of transistors. Increasing $\Delta V_{n/p}$ decreases V_{ds} of Mn1/Mp1 and so causes less leakage, just as with VAR-VDD. Because AVGS increases the source voltage, it decreases I_{SUB} and gate leakage at the same time, but in the sub-VT region I_G is not important. RBB increases the drain-bulk reverse diode current and the GIDL current because it

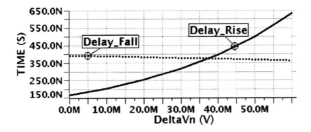

Fig. 3.21 Compensating FS corner by AVGS–PTL. Output F rise and fall delays of MUX8 gate

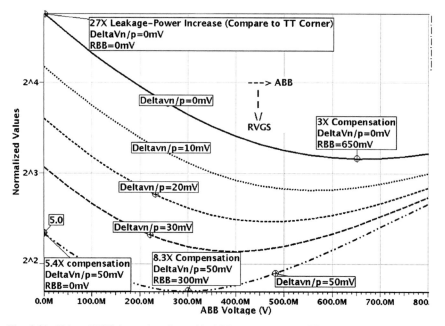

Fig. 3.22 Using AVGS in conjunction with ABB to compensate FF corner (reduced L, TOX, and VT)

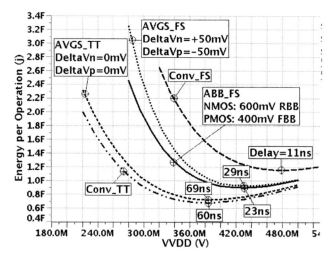

Fig. 3.23 Minimum energy point of LUT4 shown in Fig. 3.19d in TT and FS corners for both conventional (Conv) and AVGS–PTL styles. Given delay values are the critical path delay at minimum energy point

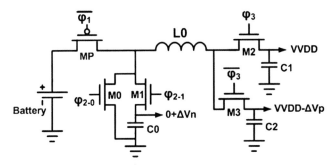

Fig. 3.24 Proposed bulk-down converter capable of generating all necessary power supplies

decreases the bulk voltage. It is important to note that every transistor is driven by $V_{gs} = \text{VVDD} - \Delta V_{n/p}$, and no transistor has $V_{gs} = \text{VVDD} - \Delta V_n - \Delta V_p$.

The area overhead is 2 to 4 transistors per gate. So in this logic style, using high fan-in gates is better. High fan-in gates also cause less leakage power due to the stack effect. In [51] it is shown that in future SOI-wafer multigate device technologies, delay increase vs. number of fan-in gates will be 41% less than bulk-CMOS. This means that high fan-in gates will work much better in these new technologies.

AVGS can be applied to bulk-CMOS, PD-SOI, FD-SOI, or FinFET. In all of these devices, increasing the source voltage when $V_g = 0$ increases the source-channel barrier height "seen" by electrons and so provides exponential control over leakage.

Figure 3.24 illustrates a modified single-inductor multioutput (SIMO) converter that can generate all of the necessary voltage levels. $\varphi1$ and $\varphi2$ are nonoverlapping. In $\varphi1$ MP is on. In $\varphi2$ one of the synchronous rectifiers, M0 or M1, can be turned on. $\varphi3$ selects which of the VVDD or $\text{VVDD} - \Delta V_p$ is to be charged. Logic gates charge C0 while L0 discharges it. With this converter it is possible to have both positive and negative $\Delta V_{n/p}$. A similar concept can be applied to switch-cap converters.

In summary, if γ-factor is sufficiently high, ABB is the best choice for compensating variations. But the technology trend shows a degradation of the body effect in all devices. In this work we proposed an AVGS method that can change the power and delay of digital gates by adjusting the V_{gs} voltage. AVGS does not need triple-well technology, works in all technology nodes, can be applied to any device, and can be used in conjunction with other conventional methods.

Conclusions

This chapter has shown some of the architectural solutions that are applicable to nanoelectronic circuits. Indeed, variability and power dissipation are two of the major hurdles that new technologies have to overcome. Crossbars realize circuits as a regular fabric and so minimize the spread of circuit delays. A new method for reducing timing variability has also been shown. Finally, ultra-low-power nanosystems require specific design technologies; adaptive V_{gs} is a very promising approach.

References

1. Holmes JD et al (2000) Control of thickness and orientation of solution-grown silicon nanowires. Science 287(5457):1471–1473
2. Luo Y et al (2002) Two-dimensional molecular electronics circuits. Chem Phys Chem 3: 519–525
3. DeHon A (2005) Design of programmable interconnect for sublithographic programmable logic arrays. In: International symposium on field-programmable gate arrays, Monterey, CA, 2005, pp 127–137
4. Choi Y-K et al (2002) A spacer patterning technology for nanoscale CMOS. IEEE Trans Electron Devices 49(3):436–441
5. Cerofolini GF (2006) Search for realistic limits to computation. II. The technological side Appl Phys A 86(1):31–42
6. Moselund KE et al (2007) Cointegration of gate-all-around MOSFETs and local silicon-on-insulator optical waveguides on bulk silicon. IEEE Trans Nanotechnol 6(1):118–125
7. Abadir MS, Reghbati HK (1983) Functional testing of semiconductor random access memories. ACM Comput Surv 15(3):175–198
8. Ben Jamaa MH et al (2008) Variability-aware design of multi-level logic decoders for nanoscale crossbar memories. IEEE Trans Comput Aided Des 27(11):2053–2067
9. Adams RD (2003) High performance memory testing. Kluwer, Dordrecht
10. Ben Jamaa MH et al (2008) A stochastic perturbative approach to design a defect-aware thresholder in the sense amplifier of crossbar memories. To appear in ASP-DAC
11. International Technology Roadmap for Semiconductors (2006) ITRS URL: http://www.itrs.net/Links/2007ITRS/Home2007.html. Accessed 12 Nov 2008
12. Strojwas A et al (1996) Manufacturability of low power CMOS technology solutions. In: Proceedings of international symposium on low power electronics and design (ISLPED), San Diego, CA, pp. 225–232
13. Eisele M, Berthold J, Schmitt-Landsiedel D, Mahnkopf R (1996) The impact of intra-die device parameter variations on path delays and on the design for yield of low voltage digital circuits. In: Proceedings of international symposium on low power electronics and design (ISLPED), San Diego, CA, pp. 237–242
14. Bowman KA, Tang X, Eble JC, Meindl JD (2000) Impact of extrinsic and intrinsic parameter fluctuations on CMOS circuit performance. IEEE J Solid State Circuits 35:1186–1193
15. Bowman K et al (2002) Impact of die-to-die and within die parameter fluctuation on the maximum clock frequency distribution for gigascale integration. IEEE J Solid State Circuits 183–190
16. Suaris P, Kgil T, Bowman KA, De V, Mudge TN (2005) Total power-optimal pipelining and parallel processing under process variations in nanometer technology. In: Proceedings of the 2005 IEEE/ACM international conference on computer-aided design (ICCAD), San Diego, CA, pp 535–540
17. Bowman KA, Austin BL, Eble JC, Tang X, Meindl JD (1999) A physical alpha-power law MOSFET model. IEEE J Solid State Circuits 34:1410–1414
18. Cao Y, Clark LT (2005) Mapping statistical process variations toward circuit performance variability: an analytical modeling approach. In: Proceedings of DAC, pp 658–663
19. Bult K (2005) Scaling effects in analog design in deep sub-micron CMOS, Lecture Notes: Advanced CMOS Circuit Design, Mead Education, 2005
20. HSPICE User's Manual (1995) Meta-Software Inc

21. Ho R et al (2001) The future of wires. IEEE Proc 89(4):490–503
22. Papoulis A, Pillai SU (2002) Probability, random variables and stochastic processes, 4th edn. McGraw-Hill, New York
23. Friedberg P et al (2005) Modeling within-die spatial correlation effects for process-design co-optimization. In: Proceedings of international symposium on quality electronic design, pp 516–521
24. Stolk PA, Widdershoven FP, Klaassen DBM (1998) Modeling statistical dopant fluctuations in MOS transistors. IEEE Trans Electron Devices 45(9):1960–1971
25. Agarwal A, Blaauw D, Zolotov V (2003) Statistical timing analysis for intra-die process variations with spatial correlations. In: Proceedings of international conference on computer aided design, pp 900–907
26. Le J, Li X, Pileggi LT (2004) STAC:statistical timing analysis with correlation. In: Proceedings of design automation conference, pp 343–348
27. Frank DJ, Solomon P, Reynolds S, Shin J (1997) Supply and threshold voltage optimization for low power design. In: Proceedings of international symposium on low power electronics and design (ISLPED), San Diego, CA, pp 317–322
28. Gronowski PE, Bowhill WJ, Preston RP, Gowan MK, Allmon RL (1998) High-performance microprocessor design. IEEE J Solid State Circuits 33:676–686
29. Hojo T (1931) Distribution of the median, quartiles and interquartile distance in samples from a normal population. Biometrika 23:315–360
30. Keeping ES (1995) Introduction to statistical inference. Dover, New York
31. Moore PG (1956) The estimation of the mean of a censored normal distribution by ordered variables. Biometrika 43:482–485
32. Chishti Z, Vijaykumar TN (2008) Optimal power/performance pipeline depth for SMT in scaled technologies. IEEE Trans Comput 69–81
33. Asenov A (2007) Simulation of statistical variability in nano MOSFETs. In: Proceedings of IEEE symposium on VLSI technology, pp 86–87
34. Das A et al (2007) Mitigating the effects of process variations: architectural approaches for improving batch performance. In: Proceedings of 34th international symposium on computer architecture (ISCA)
35. Karnik T, Borkar S, De V (2004) Probabilistic and variation-tolerant design: key to continued moore's law. Invited talk in ACM/IEEE Int'l TAU Workshop on Timing Issues
36. Yamaoka M et al (2004) Low power SRAM menu for SOC application using Yin-Yang-feedback memory cell technology. In: Proceedings of symposium on VLSI circuits, pp 288–291
37. Vittoz E (2004) Weak inversion for ultimate low-power logic. In: Piguet C (ed) Chapter 16 in Low-power electronics design, CRC Press, Boca Raton
38. Calhoun BH et al (2005) Modeling and sizing for minimum energy operation in sub-threshold circuits. JSSC 40(9):1778–1786
39. Zhai B et al (2006) A 2.60pJ/Inst subthreshold sensor processor for optimal energy efficiency. VLSI Ckts Symp
40. Verma N, Chandrakasan AP (2007) A 65nm 8T sub-VT SRAM employing sense-amplifier redundancy. In: International solid state circuits conference (ISSCC), pp 328–329
41. Kwong J et al (2008) A 65nm sub-VT microcontroller with integrated SRAM and switched-capacitor DC–DC converter. In: International solid state circuits conference (ISSCC), pp 318–319
42. Zhai B et al (2005) Analysis and mitigation of variability in subthreshold design. In: Proceedings of international symposium on low power electronics and design (ISLPED05), San Diego, CA
43. Ryan JF et al (2007) Analyzing and modeling process balance for sub-threshold circuit design. In: GLSVLSI07 Proceedings of Great Lakes Symposium on VLSI Design
44. De V et al (2001) Techniques for leakage power reduction. In: Chandrakasan A, Bowhill W, Fox F (eds) Design of high-performance microprocessor circuits. IEEE Press, Piscataway, NJ, pp 46–62

45. Calhoun BH, Chandrakasan AP (2006) Ultra-dynamic voltage scaling (UDVS) using sub-threshold operation and local voltage dithering. JSSC06 41(1):238–245
46. Tschanz J et al (2002) Adaptive body bias for reducing impacts of die-to-die and within-die parameter variations on microprocessor frequency and leakage. JSSC IEEE Journal of Solid-State Circuits vol. 37
47. Jayakumar N, Khatri SP (2005) A variation-tolerant sub-threshold design approach. In: DAC, pp 716–719
48. Singh N et al (2006) High-performance fully depleted silicon nanowire (diameter $\leq$ 5 nm) gate-all-around CMOS devices. IEEE Electron Device Lett 27(5):383–386
49. Frei James et al (2004) Body effect in Tri- and Pi-Gate SOI MOSFETs. IEEE Electron Device Lett 25(12):813–815
50. Chaudhry A, Kumar MJ (2004) Controlling short-channel effects in deep-submicron SOI MOSFETs for improved reliability: a review. IEEE Trans Device Mater Reliability 4(1): 99–109
51. von Arnim K et al (2007) A low-power multi-gate FET CMOS technology with 13.9ps inverter delay, large-scale integrated high performance digital circuits and SRAM. In: Symposium on VLSI technology digest of technical papers, pp 106–107
52. Choi J-H, Murthy J, Roy K (2007) The effect of process variation on device temperature in FinFET circuits. In: Proceedings of the 2005 IEEE/ACM international conference on computer-aided design (ICCAD), San Diego, CA, pp 747–751
53. Cho KeunHwi et al (2007) Temperature-dependent characteristics of cylindrical gate-all-around twin silicon nanowire MOSFETs (TSNWFETs). IEEE Electron Device Lett 28(12):1129–1131
54. Cho GR, Chen T (2004) Synthesis of single/dual-rail mixed PTL/static logic for low-power applications. IEEE Trans CAD Integrated Circuits Syst 23(2):229–242

Chapter 4
Actuation and Detection of Magnetic Microparticles in a Bioanalytical Microsystem with Integrated CMOS Chip

Ulrike Lehmann, Maximilian Sergio, Emile P. Dupont, Estelle Labonne, Cristiano Niclass, Edoardo Charbon, and Martin A.M. Gijs

Introduction

We present in this chapter some key technologies for lab-on-a-chip (LOC) applications, in particular related to magnetic transport and optical detection. Our objective is to show how these technologies can be realized and operate synergistically to support integrated biodetection.

Magnetic microparticles ("beads") have proven to be versatile and reliable objects in biomedical analysis and are steadily gaining interest in benchtop analytical procedures as well as in miniaturized LOC systems [1–3]. The use of micrometric particles in bioassays offers the advantage of a large specific surface for chemical binding, in combination with a high mobility imposed by the long-range magnetic forces acting on the beads. Moreover the decreasing size of LOC systems enables working with single magnetic microparticles to capture a very low number of target molecules, which evidently poses the challenge of highly sensitive detection schemes.

In this chapter, we present a hybrid microsystem that combines magnetic actuation with in situ optical detection. The chosen detection mechanism allows the observation and measurement of single magnetic microparticles of different sizes, as well as the detection of fluorescent labels attached to the particles' surface. We are able to detect mouse IgG as target antigen in a sandwich immunoassay down to a concentration of 0.1 ng/ml. Our work represents a first step toward a full diagnostic LOC system for detection of specific antigens.

Magnetic Microparticles and Magnetic Actuation

Most magnetic microparticles are made of magnetic nanocrystals enclosed in a matrix of inert and biocompatible material such as polymer or silicon dioxide

Martin A.M. Gijs (✉)
EPFL – Swiss Federal Institute of Technology, Lausanne, Switzerland

G. De Micheli et al., *Nanosystems Design and Technology*,
DOI 10.1007/978-1-4419-0255-9_4, © Springer Science+Business Media LLC 2009

[4–6]. The nanocrystals can be composed of iron oxide (maghemite or magnetite), amalgams of transition metals (Ni, Fe, Co, Mg, or Zn), or rare earth materials (NdFeB or SmCo). Depending on the interaction of the solid material with an external magnetic field, various types of magnetism can be distinguished, such as diamagnetism, paramagnetism, ferromagnetism, antiferromagnetism, and ferrimagnetism, but since magnetic microparticles should react strongly to an applied magnetic field, a high relative susceptibility χ_r is advantageous. For this reason, most magnetically responsive particles contain ferro- or ferrimagnetic material, such as maghemite or magnetite. Due to the small size of the enclosed magnetic nanocrystals, thermally induced fluctuations of the direction of magnetization cause the time-averaged magnetic moment of such nanocrystals to be zero in the absence of an external magnetic field and the bead becomes superparamagnetic [7]. In the presence of an external magnetic field, however, the magnetic moments of the crystallites tend to align with the field lines, expressing a relative magnetic susceptibility that is in the range of $\chi_r \sim 10^{-2} - 10^2$, depending on the size and the exact composition of the magnetic particle.

A superparamagnetic particle that is suspended in a liquid and is being pulled by a magnetic force experiences a set of forces, as schematically presented in Fig. 4.1. For microscopic particles, the gravitational and buoyancy forces are very small and can usually be neglected. As a consequence, the manipulation of magnetic particles in liquids is mainly guided by the two remaining diametrically opposed forces: the magnetic force that pulls on the magnetic particles and the liquid friction or drag that acts against the pull. For a moving particle in the stationary regime, we obtain

$$\vec{F}_{mag} = -\vec{F}_{friction}. \tag{4.1}$$

The magnetic force $\vec{F}_{mag}$ on a pointlike magnetic dipole can hereby generally be expressed as [8]

$$\vec{F}_{mag} = \left(\vec{m}_p \cdot \nabla \right) \vec{B}, \tag{4.2}$$

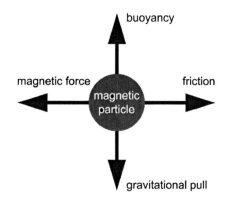

Fig. 4.1 Diagram of the forces acting on a magnetic particle in a liquid subjected to a magnetic force perpendicular to the direction of gravitation

with $\vec{m}_p$ the magnetic moment of the dipole and $\nabla\vec{B}$ the gradient of the magnetic flux density. We see from (4.2) that a nonhomogeneous magnetic field is required to obtain a translational magnetic force on the particle.

The character of the opposing liquid drag force is determined by the Reynold's number, which describes whether the flow conditions are laminar or turbulent. The limit is defined by the critical Reynold's number R_{crit}, which for water becomes $R_{crit} \approx 2100$. Due to the microfluidic systems' small size, the Reynold's number usually is much smaller than one, which puts the microsystem into the regime of Stokes flow, where viscous effects are larger than inertial forces. Therefore, the viscous drag force $\vec{F}_{drag}$ in that regime follows Stokes's law [9]:

$$\vec{F}_{drag} = -3\pi \, D_h \, \eta \, \vec{v}_s, \tag{4.3}$$

with $\vec{v}_s$ the mean velocity difference between liquid and particle, D_h the hydraulic diameter of the moving particle, and η the fluid's dynamic viscosity. The hydraulic diameter of the particle is expressed via

$$D_h = \frac{4A}{U}, \tag{4.4}$$

with A the cross sectional area and U the wetted perimeter of the cross-section.

Since the magnetic particles used in our experiments are superparamagnetic, their magnetic moment $\vec{m}_p$, for a field strength below saturation, can be expressed as follows:

$$\vec{m}_p = \int \Delta\chi\vec{H} \, dV_m, \tag{4.5}$$

with $\Delta\chi$ the difference in relative magnetic susceptibility between the magnetic material and the surrounding medium, V_m the volume of magnetic material per microparticle, and $\vec{H}$ the magnetic field. Equation (4.5) indicates the strong influence of a static magnetic field on the magnetization and thus on the magnetic force (4.2). When assuming a homogeneous susceptibility of the magnetic material over the volume V_m, (4.5) can be simplified and (4.2) transforms into

$$\vec{F}_{mag} = \Delta\chi \, V_m \left(\vec{H} \, \nabla\right) \vec{B} = \frac{1}{2\mu_0}\Delta\chi \, V_m \nabla\vec{B}^2. \tag{4.6}$$

The right-hand term is obtained via the application of the rules for the nabla operator ∇.

Based on (4.1) and (4.3), we can therefore express the velocity $\vec{v}_s$ of a magnetic particle as a function of the applied flux density $\vec{B}$ or the magnetophoretic driving force $\vec{S} \equiv \frac{\nabla B}{2\mu_0}$.

$$\vec{v}_s = \frac{1}{3\pi}\frac{\Delta\chi \, V_m}{\eta \, D_h}\frac{\nabla\vec{B}^2}{2\mu_0} = \frac{1}{3\pi}\frac{\Delta\chi \, V_m}{\eta \, D_h}\vec{S}. \tag{4.7}$$

The proportionality factor between particle velocity $\vec{v}_s$ and the magnetophoretic driving force $\vec{S}$ determines the magnetophoretic mobility m_s (a "normalized" parameter analogous to the electrophoretic mobility) of a magnetic microparticle, which depends only on properties of the particle and the surrounding medium:

$$m_s = \frac{1}{3\pi} \frac{\Delta\chi\, V_m}{\eta\, D_h}. \tag{4.8}$$

The above equation shows that particles of different sizes, whose susceptibilities differ, can express the same magnetophoretic mobility. Thus, when separating and transporting magnetic particles, neither their geometry nor their magnetic properties but a combined parameter, namely, their magnetophoretic mobility, is the relevant physical quantity.

System Setup

The magnetic manipulation system that is the focus of this chapter is based on a CMOS chip, which integrates a multilayer of microcoils with optical detection elements [10–12]. The combination of magnetic actuation and optical detection is advantageous for two reasons. Magnetic actuation offers long-range and large forces, while optical detection permits measurements of high sensitivity supported by the absence of interaction with the actuating principle. As a consequence, the manipulation and detection of single particles will be possible, which is of increasing interest for miniaturized bioanalysis. Additionally, the optical detection elements chosen on the CMOS chip enable performing on-chip fluorescent measurements, which will further enlarge the applicability of the proposed system.

The CMOS chip (Fig. 4.2) contains four metal layers, with the upper two forming the overlapping square coils, fabricated in 0.35-μm 2P4M CMOS technology. The coil double layer is responsible for the actuation of the magnetic microparticles contained in the microfluidic network placed on top of the CMOS chip. Underneath the square opening in the center of each coil, a single photon avalanche diode (SPAD) [13] is positioned, thus making the coil centers the optical detection sites.

The magnetic transport follows a three-phase current scheme [14], where an attractive coil is flanked by repulsive coils. The resulting magnetic field gradient creates a force that is sufficiently large for moving a magnetized particle to the center of the nearest attractive coil, as depicted in Fig. 4.3 [10]. In the first step, the particles are assembled around the center of the attractive coil. In the subsequent step, a neighboring coil is also changed to the attractive state, which leads to a widened but lowered magnetic field maximum, resulting in a spreading of the particle cloud over a larger area. In the third step, the first coil is changed to the repulsive state, resulting in an assembly of the magnetic particles over the neighboring, now solely attractive, coil. The optical observation of the moving cloud of particles shows that the particles follow a multitude of paths between the two coil centers, with the majority

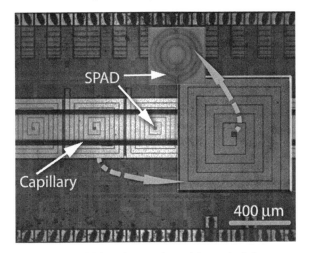

Fig. 4.2 Photograph of the CMOS-based magnetic particle manipulation system with integrated optical detection elements [single photon avalanche diodes (SPADs)]. The *insets* show a coil and a SPAD in greater detail. (Reprinted from Lehmann et al. [10], with permission © 2008 Elsevier)

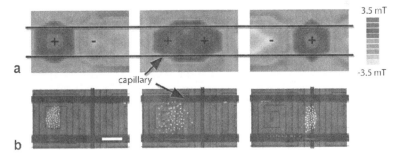

Fig. 4.3 Magnetic particle transport. **a** Simulation of the magnetic field generated by the CMOS. The "*plus*" sign demonstrates a coil in attractive mode while "*minus*" denotes a coil in repulsive state. **b** Behavior of a group of magnetic particles 5 μm in diameter according to the respective topology of the magnetic field (*dimension bar* 100 μm). (Reprinted from Lehmann et al. [10], with permission © 2008 Elsevier)

moving along a straight line. Thus the path a particle describes between the two coil centers depends on its position at the coil's center opening upon departure.

For the integrated optical detection, SPADs are chosen due to their high sensitivity, very good signal-to-noise ratio, and wide dynamic range. A SPAD is basically a p–n junction that is reverse biased above breakdown by the excess bias voltage V_e. Thus, for a SPAD in Geiger mode, every electron-creating event will trigger an avalanche and subsequently generate a countable signal. To achieve this, the voltage pulse generated during a detection cycle is regenerated and converted into a digital pulse by an inverter. Figure 4.4 shows a schematic of the SPAD and its surrounding circuit [10]. The diameter of the active region (anode) is in this case $\varnothing = 8.4\,\mu$m.

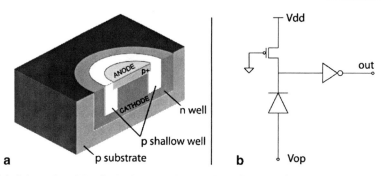

Fig. 4.4 Schematics of the single photon avalanche diode (SPAD). **a** 3D visualization of CMOS structure. **b** Electrical circuit for SPAD in Geiger mode. (Reprinted from Lehmann et al. [10], with permission © 2008 Elsevier)

In case an avalanche is triggered in this region, the resistance placed in series to the $p–n$ junction lets the reverse bias temporarily drop below breakdown and thus quenches the avalanche.

In our design, avalanche quenching, which in combination with the recharge time determines the dead time and thus the detection cycle of a SPAD, is achieved through a passive method. The measured dead time of 40 ns results in a maximum detectable frequency of 25 MHz. Thus an incident flux of 25×10^6 photons per second can be detected by the SPAD.

Optical Detection Principle

The evaluation of the signal obtained from the SPAD is based on a technique known from astronomy, where transit photometry is applied to detect and characterize extrasolar planets. This measurement principle allows the evaluation of the transiting object's properties, such as its size and the composition of its atmosphere via optical effects. In the present case, the dimensions are significantly smaller, but the ratios and observed effects are comparable. Figure 4.5a, b presents the events expected during a particle's transit. Due to the numerical aperture of the microscope's objective and the circular detection area, the light source appears as a large disc. Any particle passing between the SPAD and the light source will block a fraction of the incoming light, therefore creating a transiting dark spot or "microeclipse." Such events are known from astronomy, when planets pass between an earthbound telescope and the sun or between an observing satellite and a distant star, as shown in Fig. 4.5c. Transit photometry exploits the fact that the passage of an object (radius r) in front of a light source (radius $R \gg r$) results in a lowered light intensity measured at a detector directed toward the respective light source. Thus, the photon count of a SPAD drops during the passage of a magnetic particle between the active area and the microscope's light source.

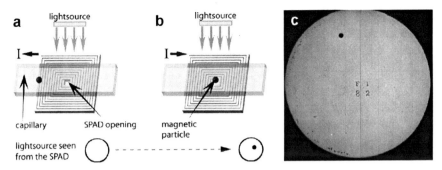

Fig. 4.5 Principle of microeclipse detection. **a** In the absence of a particle, the single photon avalanche diode (SPAD) is fully illuminated. **b** In the presence of a particle over the SPAD, the incident light is partly obstructed. **c** Transit of planet Venus on 6 December 1882. This picture, taken by American transit-expedition, is probably the oldest photograph of Venus

Experimental Results

Magnetic Particle Transport

A first range of experiments is performed in order to study the system's behavior with respect to the magnetic transport and optical detection capabilities. Magnetic particles 30 μm in diameter (Micromod) are introduced into a square glass capillary (Vitrotubes) with a bottom wall thickness of 100 μm. The capillary is then sealed and placed on the chip. The sealing of the capillary prevents the evaporation of the liquid, thereby suppressing liquid motion within the channel. Figure 4.6 shows the successful displacement of the magnetic microparticles in the system and demonstrates in addition the capability of the system to detect the presence of the particle over the SPAD at the coil center via the microeclipse effect discussed previously. Even though the particle is larger than the coil's center opening, the incident light is not fully blocked, which is in agreement with the projection model based on transit photometry.

Particle Size Measurements

In a second step, smaller particles (1, 3, 5, and 6 μm) are introduced into the system in order to examine the system's ability to distinguish between the different particle sizes. The glass capillary, used to hold the highly diluted solutions of particles, has a sidewall thickness of 25 μm, which will allow larger magnetic forces and a better resolution than the capillary used in the first experiments, due to the smaller distance between the particle and the sensing element.

Figure 4.7 shows the signal obtained from a SPAD during the transit of the different magnetic particles. The signals for particles of different sizes can be clearly

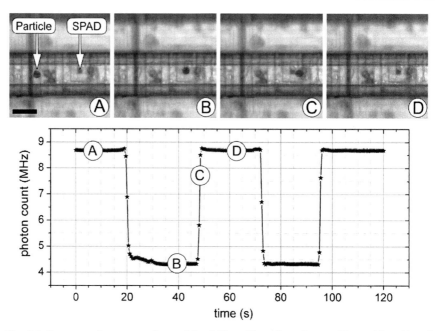

Fig. 4.6 Sequence of transport of a particle of $\varnothing = 30\,\mu\text{m}$ in a glass capillary with a sidewall $50\,\mu\text{m}$ thick in combination with the corresponding photon count of the indicated SPAD. The particle is held immobile over the coil center during the measurements (*dimension bar* $100\,\mu\text{m}$)

distinguished, and we see that particles with a diameter as low as $\varnothing = 1\,\mu\text{m}$ can be easily detected by the system. The measurements show an additional effect, which is related to the noncollimated character of the light source. Even though the phenomenon varies between the particle types, we observe that when a particle approaches the SPAD's field of vision, the photon count increases before the expected drop in intensity occurs.

This pretransition peak results from the emission, from the microscope's light source, of light of broad bandwidth and of various angles forming the objective's light cone in combination with the reflection of light falling onto the opaque particle. Thus some photons, which would normally arrive outside the SPAD's sensitive area, are diverted toward the active region and increase the measured photon flux, as schematically shown in Fig. 4.8. The probability for the redirection of photons toward the SPAD center increases as a particle gets closer to the detection area.

However, once a particle fully enters the SPAD's fields of vision, light is no longer diverted toward the SPAD's sensitive area, but away from it, which leads to the previously described effect of the microeclipse. In consequence, the number of photons arriving at the sensor surface is reduced. The evaluation of the measurements for the different particle sizes and their comparison with the model is presented in Fig. 4.9.

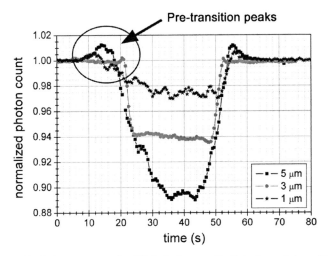

Fig. 4.7 Normalized photon count of a SPAD during transit of single particles with 1-, 3-, and 5-μm diameter. (Reprinted from Lehmann et al. [10], with permission © 2008 Elsevier)

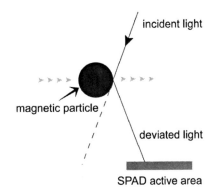

Fig. 4.8 Schematic of origin of pretransition peak. The incoming magnetic microparticle deviates incident light, arriving out of bounds, toward the SPAD's sensitive area

We see that the measurements agree well with the expected values and that, as predicted by the model, a small distance between the particles and the sensors favors high resolution. However, observation shows that, for the measurement of larger particles or cells, an increased distance is advantageous, since the upper limit of detection will increase as well. As an example, the calculated maximum diameter of a particle that can be detected in a capillary with a sidewall thickness of 25 μm is $\varnothing = 17$ μm, while a capillary with a 100-μm-thick sidewall can detect particles up to 60 μm in diameter. But we also see that the minimum size of particles that can be detected depends on the width of the capillary wall. While a capillary with a sidewall thickness of 25 μm allows the detection of particles 1 μm in diameter, a capillary with a sidewall thickness of 100 μm will yield the same signal for particles 5 μm in diameter. Thus the "working distance," defined by the channel bottom, needs to be chosen as a function of the required resolution and detection limits.

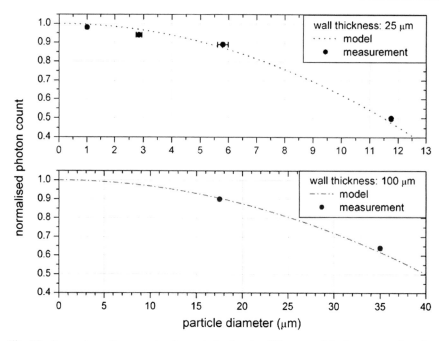

Fig. 4.9 Comparison of measured and expected values for different particle diameters and varying capillary wall thicknesses. (Reprinted from Lehmann et al. [10], with permission © 2008 Elsevier)

Particle Velocity Measurements

When recording the signals of two adjacent SPADs simultaneously, we are able to determine the average particle velocity, as Fig. 4.10 demonstrates. During the time elapsed between the intensity drops of two adjacent SPADs, the magnetic particle travels a distance of 200 μm.

The measured average velocity can be translated into the magnetophoretic mobility of the particles based on the theoretical value of the magnetophoretic driving force $\vec{S}$ in the direction of the particle displacement. Figure 4.11 shows the result of an analytical simulation of the forces along the coil centerline generated by the magnetic actuation. The magnetophoretic driving force S can be calculated using the law of Biot-Savart:

$$\vec{B} = \frac{\mu_0}{4\pi} I \int \frac{\vec{dl}}{\vec{r}^2} \times \frac{\vec{r}}{\left|\vec{r}\right|}, \tag{4.9}$$

where I is the current, r the distance between the wire and the point of interest, and $\vec{dl}$ the length of the wire section . Since each square coil consists of multiple wire segments of finite length, the field of such a coil can be determined via the superposition of the fields of all segments.

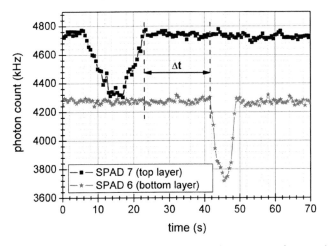

Fig. 4.10 Photon count of two adjacent SPADs during particle transport for a particle 5 μm in diameter showing the successive intensity drops. During the time difference Δt the particle moved in a straight line from SPAD 7 to the neighboring SPAD 8. (Reprinted from Lehmann et al. [10], with permission © 2008 Elsevier)

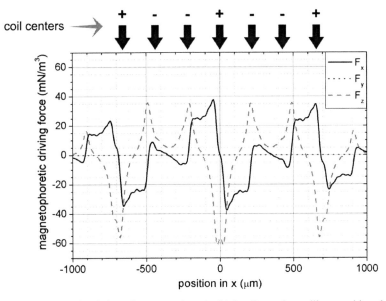

Fig. 4.11 Analytical simulation of magnetophoretic driving forces in capillary positioned over coil centers ($y = 0$). The *arrows* indicate the positions of the coil centers, with the *algebraic signs* identifying the orientation of the local magnetic field

We see that the particles experience a force in the x-direction directed from the repulsive ($-$) coils to the closest attractive ($+$) coil, while the force in the y-direction is close to zero. Additionally, the particles experience a repulsive force in the z-direction at the position of coils carrying negative currents ($-$) but are attracted

Table 4.1 Summary of average particle velocities for magnetic particles examined in system

$\varnothing(\mu m)$	Magnetite (wt%)	$v(\mu m/s)$	F_{drag} (fN)	$F_{norm}(kN/m^3)$	$\Delta\chi$
1	63.4	1.6	13.6	35.6	3.3
1.6	42.5	2.9	43.7	20.4	1.9
3	12.5	4.7	124.0	10.8	1.0
6	5.5	5.9	322.5	3.2	0.3

toward the chip when reaching the position of coils carrying positive currents (+). Thus the particles are lifted from the capillary surface and, in consequence, experience a viscous drag during the transport. When reaching the center of the next attractive coil, the particles are again pulled toward the capillary bottom, which ensures a constant distance between particle and SPAD for all optical measurements, depending only on the thickness of the capillary wall.

The comparison of the measured velocities, summarized in Table 4.1, shows a higher average velocity for the larger particles, even though their relative content of magnetically active material is lowest. The particle velocity can be translated into the viscous drag force F_{drag} acting on the particle during transport via (4.3). The obtained values are presented in Table 4.1 together with the drag force per unit volume F_{norm}.

We see that, while the velocity, and thus the magnetic force acting on the particle, increases with the diameter, the force per unit volume decreases relative to the magnetite content of the particle.

Using the relation for the magnetophoretic mobility (4.8), it is furthermore possible to determine the relative magnetic susceptibility $\Delta\chi$ of a particle. With the viscosity of the liquid of $\eta = 1\,mPa\,s$ and the average magnetophoretic driving force $S = 27\,mN/m^3$, we obtain the susceptibility values listed in the last column of Table 4.1.

When plotting the calculated relative magnetic susceptibility over the given magnetite content, we see that the magnetic susceptibility of the particles is related to the magnetite content via an exponential function (Fig. 4.12). The measured values also show the superparamagnetic character of the particles, since their relative susceptibility is significantly higher than the values known for paramagnetic materials.

Table 4.1 also demonstrates that the forces obtained by the magnetic actuation are in the fN range. For this reason, the particles can only be held against very low flow rates – on the order of micrometers per second, which makes the current system unsuitable for high flow-through bioanalytical protocols [15, 16]. The introduction of an additional retention system, e.g., via dedicated coils, could solve this problem and thus opens the way toward stop-flow bioanalytical protocols using single magnetic particles or magnetically labeled cells.

However, the system's capability for the optical measurements of particle size and particle velocity will permit determining the magnetic parameters of particles or magnetically labeled cells. Especially in the case of cells, the magnetophoretic mobility can be used as a measure for the number of magnetic particles attached to

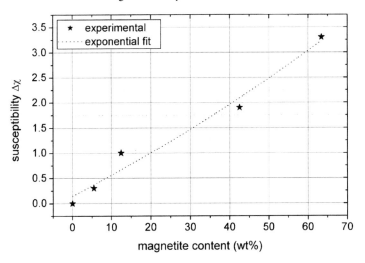

Fig. 4.12 Magnetic susceptibility $\Delta\chi$ as obtained from velocity measurements for known particle sizes

the cell surface, which are in turn an indicator for the number of specific binding sites expressed by the cell [17–19].

Fluorescent Measurements

The particle size measurements demonstrate the high sensitivity of the optical detection system, which is able to detect the intensity drop created by the passage of a particle with a diameter of 1 μm. Additionally, it is known that SPADs are able to perform fluorescent measurements. Thus by replacing the broadband light source with a UV lamp (X-Cite series 120, Zeiss), the system can be easily changed into a configuration for on-chip fluorescent detection. For an efficient fluorescent measurement, it is thus important that the emission wavelength matches the maximum sensitivity of the SPAD.

Based on standard immunoassay protocols (Fig. 4.13), we can envision a LOC that allows the detection of a fluorescent label captured at the surface of a single magnetic particle. In order to test the fluorescent detection principle, we first use nonfluorescent streptavidin-coated magnetic particles 3 μm in diameter, to whose surface biotinylated antibodies with a fluorescein isothiocyanate (FITC) label can be bound. An estimation of the particles' loading capacity, according to the information given by the supplier, yields 2.6×10^6 antibodies that can be bound to the particle surface until saturation point is reached. In terms of total weight, this amount of antibodies corresponds to 0.6 pg of bound material.

When a particle that carries fluorescent molecules at its surface approaches the coil center, the photon count at the sensor will drop due to the microeclipse effect,

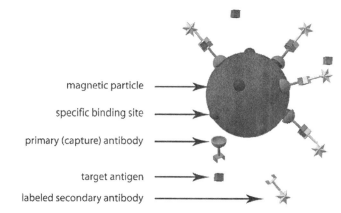

Fig. 4.13 Schematic of a sandwich immunosorbent assay on a microparticle. Fluorescent labels allow the quantification of the captured sample

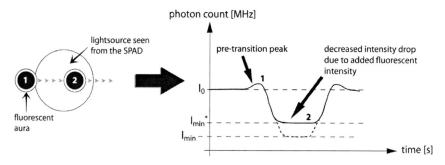

Fig. 4.14 Principle of fluorescent detection. The fluorescent aura around the particle leads to an increase in the light intensity measured during the microeclipse

but additionally the fluorescent circumference will influence the signal under suitable light conditions. Figure 4.14 demonstrates the case where the system is set to the fluorescence mode. Due to the light emission at the particle's circumference, the photon count of the SPAD will be higher than in the case of a nonfluorescent particle. The difference between the two signals is the indication for the presence of fluorescent molecules at the particle surface. Such a differential measurement can be performed by varying the wavelength of the light between two measurements on the same particle, which allows recording the microeclipse signal of the same particle with and without the fluorescent component.

The origin of the fluorescent circumference is schematically presented in Fig. 4.15, which depicts how an incident light wave is reflected at the particle surface and is deviated from its path toward the sensor surface. While passing the zone of the fluorescent molecules at the particle surface, the incident or reflected light excites a molecule, which will emit light in any possible direction. Thus a certain fraction of the light emitted in the fluorescently active zone at the particle surface

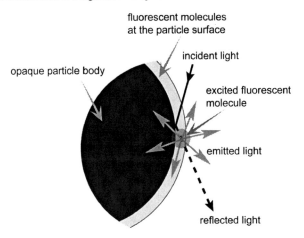

Fig. 4.15 Schematics of formation of fluorescent aura. The undirected emission of light from molecules excited by blocked or deviated light waves creates an additional light source at the particle circumference

will arrive at the sensor and increase the amount of counted photons. The part of the light-emitting zone "seen" by the SPAD is limited to a stripe along the particle circumference, since the opaque particle body blocks the light emitted at the upper particle hemisphere and prevents the lower hemisphere from being illuminated. In consequence, the measured fluorescent signal does not correspond directly to the total amount of the captured fluorescent molecules but is an indicator of the density of the fluorescent molecules at the particle surface, since a decrease in density will result in a smaller number of active molecules at the particle's circumference. For example, the density of antibodies at the surface of a fully saturated particle 3 μm in diameter amounts to 8.6×10^4 molecules/μm^2, which presents the upper limit of detection.

Figure 4.16 shows the SPAD signals of a fully saturated particle illuminated at the excitation wavelength, resulting in a fluorescent emission, and at a different wavelength, resulting in a nonfluorescent microeclipse signal. The drop in the photon count is significantly smaller when the fluorescent emission occurs, which indicates the expected presence of an additional photon source. We further notice that the pretransition peaks observed for the broadband light source also occur during the fluorescent measurements but are clearly smaller than the signal change observed in the experiments employing a broadband light source.

Measurements show that the difference between the nonfluorescent particle and the particle fully covered with fluorescent molecules is on average 66% of the total signal change, which is equivalent to the maximum density of bound antibodies and therefore stands for the maximum fluorescent signal.

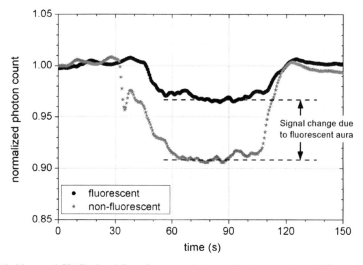

Fig. 4.16 Measured SPAD signal for a 3-μm particle at excitation wavelength (fluorescent) and at a nonexciting wavelength (nonfluorescent)

Bioanalytical Experiments

In a subsequent development, we look for a correlation between varying concentrations of the target antigen in a sandwich immunoassay protocol (Fig. 4.13) and the differential signal readout from the SPAD. Beads with a mean diameter of 2.9 μm (Bangs Labs) are used as mobile substrate. These beads are coated with streptavidin. A first capture antibody is grafted to the beads, by mixing for 10 min 10 μl of beads, diluted in 100 μl of immobilization buffer, with 100 μl of high-concentration (50 μg/ml), biotinylated, polyclonal rabbit antimouse IgG (Dako). Then, beads are retained by means of a magnet and washed for 30 s in 500 μl of a phosphate buffer saline–bovine serum albumin (PBS–BSA) solution. The next step is mixing for 10 min the beads with 100 μl of solution in which different concentrations of the target antigen (the analyte) are dissolved. After a washing step with PBS–BSA, a further step consists in mixing for 10 min the functionalized beads with 100 μl of solution at 167 μg/ml of Cy3 conjugated affinity purified goat antimouse IgG (Rockland). The last step is washing the beads three times in PBS–BSA solution (500 μl each time).

When reducing the target antigen concentration, the coverage of bound fluorescent labels is reduced, leading to a decrease in the fluorescent signal, as Fig. 4.17 clearly shows. Antigen concentrations above 100 ng/ml lead to a saturation of the response. This result indicates that the particle is fully covered by the sample molecules at this concentration. The results obtained for lower concentrations clearly show the expected decrease in the fluorescent signal. The high sensitivity of the SPADs allows us to detect concentrations as low as 66 pg/ml of analyte.

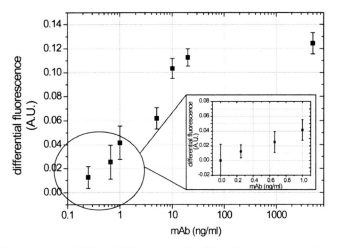

Fig. 4.17 Measurement of differential fluorescent signal for different concentrations of the IgG monoclonal antibody (mAb) that is the target antigen of our test. The *error bars* indicate the variability in fluorescent response for different magnetic beads

A further remaining challenge is the incubation of the magnetic particles on-chip and performing the sandwich immunoassay fully on-chip. Indeed, the measurements presented above have all been performed using particles prepared and incubated off-chip. Since this type of procedure still requires larger amounts of particles and reagents than actually needed for the on-chip detection, the successful demonstration of the fluorescent detection capabilities of the magneto-optical CMOS system will logically be combined with on-chip retention and manipulation schemes for the magnetic microparticles.

Conclusion

This chapter successfully demonstrated the capabilities of a magneto-optical CMOS microsystem for the on-chip detection of biological molecules. Our results show the potential of such a system with respect to on-chip bioanalysis. We were able to show that a miniaturized magnetic transport system could be combined with highly sensitive optical detectors, which allows the direct link between magnetic actuation and nonconfocal optical detection. The high sensitivity of the detection allows a differentiation between objects of different sizes as well as varying load of fluorescent molecules. Since it is equally possible to measure the velocity of transport between the detection sites, the magnetic properties of the handled objects can be determined as well. In summary, our magneto-optical CMOS system presents a promising and innovative approach to the design of highly sensitive and flexible bioanalytical microsystems and is anticipated to be able to handle and detect a wide range of biomolecules and cells.

References

1. Gijs MAM (2004) Magnetic bead handling on-chip: new opportunities for analytical applications. Microfluid Nanofluidics 1:22–40
2. Pankhurst QA, Connolly J, Jones SK, Dobson J (2003) Applications of magnetic nanoparticles in biomedicine. J Phys D Appl Phys 36:R167–R181
3. Pamme N (2006) Magnetism and microfluidics. Lab Chip 6:24–38
4. Landfester K, Ramirez LP (2003) Encapsulated magnetite particles for biomedical application. J Phys Condens Matter 15:S1345–S1361
5. Bergemann C, Muller-Schulte D, Oster J, a Brassard L, Lubbe AS (1999) Magnetic ion-exchange nano- and microparticles for medical, biochemical and molecular biological applications. J Magn Magn Mater 194:45–52
6. Grüttner C, Rudershausen S, Teller J (2001) Improved properties of magnetic particles by combination of different polymer materials as particle matrix. J Magn Magn Mater 225:1–7
7. LesliePelecky DL, Rieke RD (1996) Magnetic properties of nanostructured materials. Chem Mater 8:1770–1783
8. Zborowski M, Sun LP, Moore LR, Williams PS, Chalmers JJ (1999) Continuous cell separation using novel magnetic quadrupole flow sorter. J Magn Magn Mater 194:224–230
9. White FM (1999) Fluid mechanics. McGraw-Hill, New York
10. Lehmann U, Sergio M, Pietrocola S, Dupont E, Niclass C, Gijs MAM, Charbon E (2008) Microparticle photometry in a CMOS microsystem combining magnetic actuation and in-situ optical detection. Sens Actuators B Chem 132:411–417
11. Lehmann U, Sergio M, Pietrocola S, Niclass C, Charbon E, Gijs MAM (2007) A CMOS microsystem for the integrated actuation and optical detection of magnetic microparticles. The 14th International Conference on Solid-State Sensors, Actuators and Microsystems, Transducers'07 and Eurosensors XXI, pp U1255–U1256, Lyon, France
12. Lehmann U, Sergio M, Pietrocola S, Niclass C, Gijs MAM, Charbon E (2007) Particle shadow tracking – combining magnetic particle manipulation with in-situ optical detection in a CMOS system. The 11th International Conference on Miniaturized Systems for Chemistry and Life Sciences, μTAS 2007 Conference, pp 1179–1181, Paris, France
13. Niclass C, Sergio M, Charbon E (2006) A single photon avalanche diode array fabricated in 0.35 mu m CMOS and based on an event-driven readout for TCSPC experiments. Conference on Advanced Photon Counting Techniques, pp U216–U227, Boston, MA
14. Rida A, Fernandez V, Gijs MAM (2003) Long-range transport of magnetic microbeads using simple planar coils placed in a uniform magnetostatic field. Appl Phys Lett 83:2396–2398
15. Lacharme F, Vandevyver C, Gijs MAM (2008) Magnetic bead retention device for full on-chip sandwich immuno-assay. 21st IEEE International Conference on Micro Electro Mechanical Systems MEMS 2008, pp 184–187, Tucson, AZ, USA
16. Lacharme F, Vandevyver C, Gijs MAM (2008) Full on-chip nanoliter immunoassay by geometrical magnetic trapping of nanoparticle chains. Anal Chem 80:2905–2910
17. Yi CQ, Li CW, Ji SL, Yang MS (2006) Microfluidics technology for manipulation and analysis of biological cells. Anal Chim Acta 560:1–23
18. McCloskey KE, Chalmers JJ, Zborowski M (2003) Magnetic cell separation: characterization of magnetophoretic mobility. Anal Chem 75:6868–6874
19. McCloskey KE, Moore LR, Hoyos M, Rodriguez A, Chalmers JJ, Zborowski M (2003) Magnetophoretic cell sorting is a function of antibody binding capacity. Biotechnol Prog 19:899–907

Chapter 5
Thin Film Bulk Acoustic Wave Resonators for Gravimetric Sensing

Evgeny Milyutin and Paul Muralt

Introduction

Gravimetric sensing with a piezoelectric oscillator subject to mass loading is a proven technique to measure small mass changes and has been used routinely for decades in thin film deposition [1,2]. Later, this technique was also widely explored for use in sensors measuring accumulation of chemical or biological species [3,4]. The principle is depicted in Fig. 5.1. The resonator is coated with a chemically active layer immobilizing biological or chemical species. In the early years, the piezoelectric oscillator was based on a resonating AT-cut quartz plate [3] whose frequency was defined by the thickness of the plate. In such a device – based on a bulk wave – the acoustic energy is mainly distributed inside the plate. Logically, the sensitivity with respect to surface effects will increase by thinning down the plate. This leads, however, to quite brittle structures. Further solutions were investigated, targeting waves localized in a surface layer: surface acoustic waves [5], surface transverse waves (see in [4]), and Love waves [6, 7]. Furthermore, membrane-type structures with Lamb wave plate modes [5] and, more recently, thin film bulk acoustic resonators (TFBARs) [8] were investigated. However, they also resulted in brittle membrane structures consisting of piezoelectric thin films (ZnO, AlN) on thin elastic layers (Si, Si_3N_4, etc.). TFBARs allow for avoiding the brittle membranes by the use of acoustic Bragg reflectors. For this reason, such TFBARs were introduced as "solidly mounted resonators" (SMRs) when proposed for RF-filter technology [9]. A number of studies with SMR sensors have been carried out in recent years [10–13].

An important differentiation of the various acoustic excitation mechanisms is the ability to create exclusively shear displacements, meaning that at the surface, the displacement is tangential, as shown in Fig. 5.1. This property is required for operations in liquids because shear modes do not emit waves into liquids and thus are able to maintain a high quality factor in immersed operation. Such shear mode sensors are required for the largest potential markets, which are medical diagnosis,

Evgeny Milyutin (✉)
EPFL – Swiss Federal Institute of Technology, Lausanne, Switzerland

G. De Micheli et al., *Nanosystems Design and Technology*,
DOI 10.1007/978-1-4419-0255-9_5, © Springer Science+Business Media LLC 2009

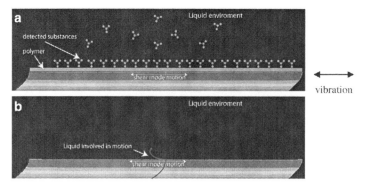

Fig. 5.1 **a** Immobilization of molecules by gravimetric sensor depicted in a schematic cross-section. **b** Schematic cross-section through acoustically active plate in shear mode oscillation in a liquid, as obtained by finite element modeling. The motion amplitude decays in the liquid, according to its viscosity

drug screening, and, eventually, environmental sensors. The surface of the device vibrates in the transverse direction. An organic (polymer) layer acting as biochemical affinity layer, or immobilization layer, is deposited on top, possibly separated by an acoustic impedance matching layer. The whole is immersed in a liquid (mostly water) and exposed to biological species, molecules, or ions to be detected. The liquid is dragged with the surface to a depth defined by its viscosity (Fig. 5.1b). The interaction with the captured molecules is quite complex and involves not only their mass and density but also changes in viscosity in the adjacent liquid and changes in the rigidity of the polymer layer.

The interesting feature of piezoelectric oscillators is their ability to be used as transducers, meaning that the device is at the same time an actuator for wave excitation (converse piezoelectric effect) and sensor to evaluate the response (direct piezoelectric effect). This allows for compact designs. A frequency scan allows quick assessment of the resonance frequency as the mechanical resonance shows up in the electrical impedance or admittance spectrum. The sensor output is the measured resonance frequency shift due to the mass change. Figure 5.2 shows an example of an 8-GHz SMR resonator loaded with a self-assembled monolayer.

If we assign the frequency shift to mass loading only, it is customary to define a mass sensitivity as follows (see, e.g., [4]):

$$S_{\mathrm{m}} = \lim_{\Delta m \to 0} \frac{1}{f} \frac{\Delta f}{\Delta m}. \tag{5.1}$$

For small mass changes Δm, the relative frequency change is then given by

$$\frac{\Delta f}{f} = S_{\mathrm{m}} \Delta m. \tag{5.2}$$

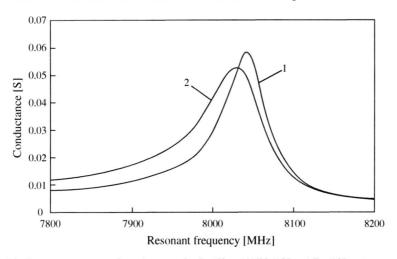

Fig. 5.2 Frequency spectra of conductance for Pt (60 nm)/AlN (180 nm)/Pt (100 nm) resonator. The spectrum (1) was made in air with an uncoated resonator and the spectrum (2) with a resonator covered with a self assembled monolayer 11-mercaptoundecanoic acid attaching on the top electrode Pt film

It is intuitively clear that the smaller the mass of the oscillator, the higher the sensitivity must be. Indeed, if we approximate the oscillator by a simple spring whose angular resonance frequency is given by $\omega_0 = (k_s/m)^{1/2}$, we obtain a mass sensitivity of

$$S_m \text{ (spring)} = \frac{1}{\omega_0} \frac{\partial \omega}{\partial m} = -\frac{1}{2\omega_0} \frac{k_s^{1/2}}{m^{3/2}} = -\frac{1}{2m} \stackrel{(1)}{=} -\frac{\omega_0^2}{2k_s} \stackrel{(2)}{=} -\frac{\omega_0}{2 (k_s m)^{1/2}}, \quad (5.3)$$

which yields a sensitivity proportional to $1/m$. This relation can be generalized to other types of resonators (SAW, BAW, Lamb) yielding the same factor $1/(2m)$ as in (5.3) [14], where m is the effective mass in motion in the given resonator. The frequency dependence of the sensitivity is, however, not easily derived in the general case. In the simple example of an oscillating spring, the mass is independent of the spring constant (when we can neglect the mass of the spring as compared to the attached mass), and the sensitivity should increase as the square of the frequency [relation (1) in (5.3)]. However, ultrasonic wave propagation depends on mass density and stiffness at the same time, and eventually the sensitivity increases only linearly with frequency, as suggested by the second equality in (5.3). A more detailed analysis will be given later in this chapter.

Alternatively, the sensitivity is also defined as relative frequency shift per change of surface mass density μ (unit kg/m^2), which yields a number that is independent of the sensor area:

$$S_\mu = \frac{1}{f_0} \lim_{\mu \to 0} \frac{|\Delta f|}{\mu}. \quad (5.4)$$

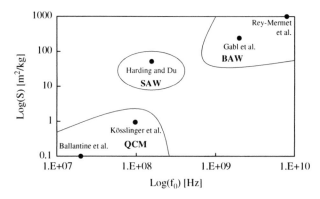

Fig. 5.3 Overview of sensitivity of mass density obtained with quartz micro balances (QCM) [15, 16], SAW devices [17], and BAW devices [10, 12] following the sensitivity definition of (5.4)

Comparing published data from a number of works, a rather linear increase in sensitivity with frequency is indeed observed (Fig. 5.3). It is thus interesting to explore gravimetric sensors working at higher frequencies. With the advent of MEMS technology and progress in the processing of piezoelectric thin films, it became possible to fabricate BAW resonators based on tiny sheets of such films and to raise the frequency into the gigahertz range. The goal of this article is to give an overview on recent trends in this field and to present in more detail a novel type of shear mode SMR based on interdigitated electrodes (IDEs).

Piezoelectric Transducers Based on Thin Films

General Considerations on Growth, Electrodes, and Operation Modes

Introductory texts on thin film piezoelectrics and MEMS can be found elsewhere [18, 19]. In this article we deal exclusively with the nonferroelectric polar materials ZnO and AlN growing in the hexagonal wurtzite structure. The growth process must not only provide for a film texture, but also for the alignment of the polar direction (c-axis in this case). This is indeed possible, even at low temperatures of 200 to 300°C when using reactive sputter deposition. The specific mechanism of this technique to allow for excellent piezoelectric films seems to be related to ion bombardment providing sufficient surface mobility of ad-atoms. From the various literature available on the issue [20–22], we can make a general conclusion as follows: In a certain range of ion energies between a lower limit for affecting ad-atom mobility (about 20 eV) and a higher limit given by a threshold above which implantation into the first sublayer becomes important (about 70 eV), there is an optimal ion impact providing excellent c-axis orientation with defined polarity. Deviations from perfect c-axis orientation are possible when during the nucleation stage tilted grains are allowed to grow. Different orientations are obtained at lower ion bombardment energies [22] or with a higher surface roughness of the growth substrate [23]. The growth of tilted grains requires, firstly, the nucleation of grains that are not (001)-oriented and, furthermore, an oblique flux of the arriving atoms from the sputter target in order to give a growth advantage for non-(001)-oriented grains [24, 25] such as (101) or (102), etc. It's not evident how to achieve a homogeneous film on a large wafer. By tilting the wafer away from the usual face target position, the deposition uniformity becomes very bad (no wafer rotation is possible).

Considering a (001)-textured, polycrystalline, columnar film slab as a free body, we would find a thickness mode d_{33} that is equal to the $\delta_{3'3'}$ of the single crystal column (3': crystal axes). The in-plane ordering is random, resulting in a cylindrical symmetry. This gives the same symmetry of the piezoelectric tensor as for hexagonal or tetragonal crystal symmetry. It follows that in the case of wurtzites, d_{31} is equal d_{32} and corresponds to the crystal $\delta_{3'1'}$ value. In addition, the shear mode coefficients d_{15} and d_{24} coincide as well and again correspond to the single crystal value. Any electric field in the film plane – thus perpendicular to direction 3 along which the polar axes of the grains are pointing – is thus able to excite a shear mode in the film as in a single crystal. Some standard electrode configurations are schematically shown in Fig. 5.4.

The piezoelectric coefficient of a tilted single crystal is readily obtained from coefficients defined for the single crystal directions ($\delta_{3'3'}$, $\delta_{3'1'}$, $\delta_{1'5'}$). Knowing the dielectric tensor, as well as the rigidity tensor, the coupling coefficient k_t^2 for a (clamped) plate in thickness mode BAW resonance, excited with a parallel plate capacitor structure, can be calculated (Fig. 5.5). The highest coupling coefficient is obtained for a tilt of around 50°. In practice, the tilt angle amounts to around 20 to 30° [26].

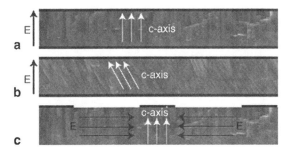

Fig. 5.4 Electrode systems for driving piezoelectric films. **a** Planar capacitor structure with top and bottom electrode with c-axis oriented polar film. **b** Planar capacitor structure with tilted c-axis-oriented polar film. **c** c-Axis-oriented polar film grown on insulator and having interdigitated electrode on top, enabling an in-plane electric field between electrode fingers. This configuration is usually used for SAW excitation

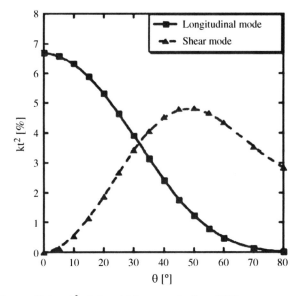

Fig. 5.5 Coupling coefficient k_t^2 of plate thickness mode (from [26])

Shear Mode with C-Axis-Oriented AlN and In-Plane Electric Field

In this section we consider the situation sketched in Fig. 5.4c for the excitation of a shear bulk wave. We show that such a wave can be trapped in the piezoelectric film slab. The latter (for instance made of AlN) consists of grains having their polar c-axis everywhere perpendicular to the plane of the plate (along direction 3). The simple situation with the electric field in the film plane (direction 1) can be solved analytically. This electromechanical problem requires the simultaneous application of Newton's wave equation for the mechanical wave part and Maxwell's equation of the electrical part [relativistic effects requiring in addition a magnetic field are neglected (i.e., $B = 0$)]. The equations of state of the piezoelectric contain a coupling term between electrical parameters E and D, and the mechanical parameters stress (T) and strain (S). In the following notation, the piezoelectric coefficient is the e-type relating stress to the electric field, or displacement field to strain. The resulting equations read as follows:

$$\rho \ddot{u}_i = \frac{\partial T_{ij}}{\partial x_j}, \tag{5.5}$$

$$T_{ij} = c_{ijkl}^E S_{kl} - e_{nij} E_n,$$

$$D_n = \varepsilon_{nj} E_j + e_{nkl} E_n, \tag{5.6}$$

$$\text{rot} \vec{E} = 0,$$

$$\text{div} \vec{D} = 0, \tag{5.7}$$

where u_i, T_{ij}, S_{ij}, E_j, and D_n are mechanical displacement field, stress and strain tensors, electric field intensity, and electric displacement field, respectively. The upper index in c_{ijkl}^E denotes that the condition at a constant electric field. In what follows, we will use a reduced index notation and understand c_{ij} to stand for the constant E-field. In the applied coordinate system, direction 3 points perpendicular to the film plane and is at the same time the 6-fold polar axis of the AlN single crystal grain.

For an electric field pointing along the x_1-direction (in-plane direction), (5.3) becomes

$$T_i = c_{ij} S_j, \tag{5.8}$$

where $i, j = 1, 2, 3, 4, 6$ and

$$T_5 = c_{55} S_5 - e_{15} E_1. \tag{5.9}$$

From the last equation one can see that only the S_5 strain of the film is coupled with the electric field E_1 and that this deformation is not coupled with any other deformation. Hence the excited wave contains only displacements corresponding to S_5. For a clamped thin film, S_5 is equal to $\frac{\partial u_1}{\partial x_3}$. Introducing Maxwell's equations, it follows that all variables depend on x_3 only. Then by solving Maxwell's equations and Newton's equation together with the equations of electromechanical coupling we obtain

$$\frac{\partial E_1}{\partial x_3} = 0, \frac{\partial E_1}{\partial x_1} = 0, \tag{5.10}$$

$$\rho \ddot{u}_1 = \frac{\partial T_5}{\partial x_3}, \tag{5.11}$$

$$\frac{\partial T_5}{\partial x_3} = c_{55} \frac{\partial S_5}{\partial x_3} = c_{55} \frac{\partial^2 u_1}{\partial x_3^2}, \tag{5.12}$$

$$\rho \ddot{u}_1 = c_{55} \frac{\partial^2 u_1}{\partial x_3^2}. \tag{5.13}$$

The last equation describes shear waves that propagate in the x_3-direction [wave vector $(0, 0, x_3)$] and with displacement in the x_1-direction. The current and voltage between electrodes can easily be derived and admittance $Y = I/U$ is finally equal to

$$Y = j\omega C_0 \left(1 + \frac{e_{15}^2}{c_{55}\varepsilon_{11}} \frac{\tan(\varphi)}{\varphi} \right), \varphi = \frac{k_3 l}{2}, \tag{5.14}$$

where C_0 is the static capacitance between the electrodes and wave number $k_3 = \omega \sqrt{\frac{\rho}{c_{55}}}$. Resonances are observed for infinite values of $\tan(\varphi)$. At the antiresonance frequency the admittance becomes zero. The difference between the resonance and antiresonance frequency is proportional to the piezoelectric coupling constant

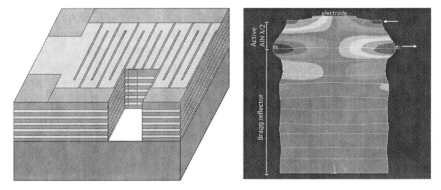

Fig. 5.6 Schematic drawing of SMR shear mode BAW device with interdigitated electrodes (*left*), indicating the cut for modeled section. The resulting transverse displacements (*right*) at the basic resonance were obtained by finite element modeling, including a boundary element method subprogram to deal with the thick substrate

$k_{15}^2 = \frac{e_{15}^2}{c_{55}\varepsilon_{11}}$. Note that the value of e_{15} does not affect the resonance frequency but strongly affects the antiresonance frequency. This is a result of the fact that c_{55} at constant E-field is relevant for wave propagation, as follows from Maxwell's equations (5.3).

In practice, it is difficult to create a pure in-plane electric field because it would require very precise patterning of the active AlN film without avoiding completely stray fields. A more practical way of supplying in-plane electric fields is the use of IDEs, as shown in Fig. 5.4c. In this geometry, the electric field has opposite directions in neighboring IDE sections leading to an antiphase motion in adjacent regions. Modeling by a finite element method with a boundary element method (FEM–BEM) can be used to describe the resulting motion of such a structure.

The properties of the SMR with a periodic planar electrode system were simulated with the help of ANSYS. A half period of the mechanical response at resonance frequency is shown in Fig. 5.6. The electrode is located in the center at the top of the structure, the bottom border is clamped to a semi-infinite substrate, and half-period conditions are established along the horizontal axis. Bragg's reflector is located between the substrate and the resonating AlN film in order to provide acoustic isolation. The fact that we are dealing with shear waves can easily be ascertained from the type of resonant motion within the active AlN layer.

It is important to emphasize several features of the wave. (1) A shear mode thickness resonance can be seen between the electrodes (to the left and right of the center of the structure). (2) The most intense motion is concentrated in the top layer. The amplitude decreases strongly within the Bragg reflector. (3) The displacement of the AlN surface between electrodes is mostly horizontal, which is very important for sensors in liquid applications. (4) There are small vertical displacements of the electrode, which are expected to contribute to losses in immersed operation.

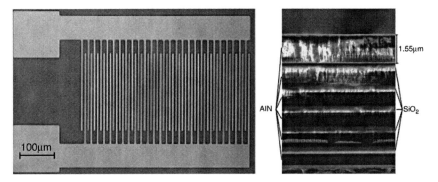

Fig. 5.7 *Top view* and TEM cross-section of a realize shear mode device with interdigitated electrodes (from [13])

Realized Shear Mode Resonator with C-Axis-Oriented AlN Thion Films

A top view and a cross-sectional view of a fabricated device are shown in Fig. 5.7. The device consists of a reflector composed of five pairs of SiO_2 and AlN layers. Such reflectors were originally proposed and fabricated by Lakin et al. for RF filters based on the longitudinal mode [9]. They require more layers than the commonly used SiO_2/W type [27] but have the advantage of being electrically insulating. This is extremely important for IDE shear mode excitation. Thin films of AlN were deposited by pulsed dc reactive magnetron sputtering, and the SiO_2 films by rf-sputtering at a temperature of 300°C. The active AlN layer is deposited on top of the last SiO_2 layer of the reflector. A very thin layer of SiO_2 on top of the active AlN layer was used to protect the AlN layer from being attacked by developer solution during subsequent photolithography for aluminum electrode patterning. The 150-nm-thick aluminum electrodes with chromium adhesion layers were evaporated at room temperature and patterned by means of a lift-off process. This device thus needs only one photolithography step, and not two like the simplest possible TFBAR.

All devices were designed to have equal static capacitance, but at the same time the distance between the centers of neighboring electrodes' fingers was varied from 6 to 10 μm in order to test the resonance frequency dependence on the IDE dimensions. The resonance frequency was found to change very weakly with increasing distance between adjacent finger electrodes. The change amounted to less than 2% when the distance between fingers was almost doubled from 6 to 10 μm. The devices for this study were built near each other on the wafer, and thus all layers had equal thickness. The thickness mode nature of the resonance is thus confirmed, excluding any waves dependent on IDE periodicity [13].

Figure 5.8 shows a measured real part of admittance (conductance). The quality factor Q, which is evaluated as the ratio between full width at half maximum of the conductance peak, was determined to be 870 in air. The conductance curve of the

Fig. 5.8 Real part of admittance for a shear SMR operated in air and in liquid (silicon oil) (from [13])

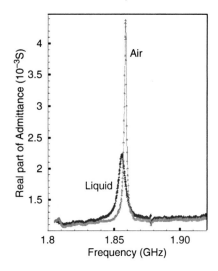

device immersed in silicon oil is shown in the same figure. The peak amplitude is decreased, as is the Q-factor, and a small shift of the resonance frequency (2 MHz) takes place. This shift may have two possible origins. A first one might be a loading effect due to the local piston motion of the electrode regions, which contain the previously discussed longitudinal (i.e., vertical) wave component. A second origin could be surface roughness or asperities. As the liquid has some finite viscosity, it is dragged to a certain depth, and thus the shear wave is loaded as well. The Q-factor measured in immersed operation amounted to 260. In comparison, Wingqvist et al. [8] reported a Q-factor of 150 that was achieved by a shear mode TFBAR based on inclined c-axis AlN and operated at a lower frequency of 1.2 GHz.

Realized Gravimetric Thin Film BAW Sensors and Their Performance

TFBAR or SMR gravimetric sensors were first realized in versions operating with longitudinal waves [10, 12, 28]. The increase in sensitivity with frequency is indeed observed, as shown in Fig. 5.3, which depicts the sensitivity reported in selected works [12]. The BAW numbers given in this figure correspond to longitudinal modes. Such devices could be useful as gas detectors. A thin polymer layer on the resonator would serve as a gas-sensitive layer changing its mass and its elastic constant with absorption of the gas.

As discussed above, detection in a liquid as required for biomedical sensors requires the use of shear mode BAWs. The resonator is less damped in liquids, and a good resolution is still possible [8,11,24]. The most advanced prototype sensor is the one from the Siemens group [11] (Fig. 5.9). It is based on 18° tilted ZnO with a parallel plate capacitor structure (as in Fig. 5.3b) resonating at 790 MHz. The reflector

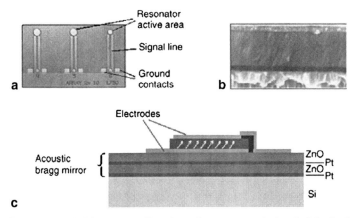

Fig. 5.9 Setup of FBAR biosensor. **a** *Top view* of resonator and electrical leads. **b** Electron microscope picture of c-axis-inclined ZnO. **c** Schematic picture of lateral structure comprising the resonator with two electrodes solidly mounted on an acoustic Bragg mirror (from [11])

is realized as a ZnO/Pt multilayer stack. This sensor was tested in an immunoassay using avidin/antiavidin detection as an example. Sensitivity and signal-to-noise (SN) were evaluated for the SMR as well as for a QCM reference sensor. The sensitivity of the first was $800\,Hz\,cm^2/ng$, corresponding to $100\,m^2/kg$ when dividing by the resonance frequency to follow the definition in (5.4). The reference 10 MHz QCM yielded a sensitivity of $0.22\,Hz\,cm^2/ng$, or $2.2\,m^2/kg$. These values are quite compatible with the sensitivity plotted in Fig. 5.3. However, differences in quality factors (100 to 150 for TFBAR and 2,000 for QCM) as well as higher noise levels increased the detection limit of TFBAR. Still, the detection limit of TFBAR was twice as much $-2.3\,ng/cm^2$ vs. $5.2\,ng/cm^2$ for QCM.

Wingqvist et al. have fabricated a biochemical sensor based on inclined c-axis AlN for cocaine and heroin detection. The device was incorporated into a micromachined fluidic system. A comparative study of QCM and TFBARs showed that the integrated fluidic system is an important element of the sensor design inasmuch as the small size of the TFBAR can complicate the liquid delivery handling.

The work of the Siemens group raises the question of the relation between sensitivity and SN ratio. In this work, the sensitivity was increased by a factor of 50; the SN ratio, however, increased by a factor of 2 only. This is a success, no doubt; however, the question remains as to how we must understand that small improvement with frequency increase. Analyzing in more detail the sensitivity as given in (5.4)

$$S_\mu = \frac{1}{f_0} \lim_{\mu \to 0} \frac{|\Delta f|}{\mu}.$$

we consider the simple case when loading with a material of identical acoustic properties as the piezoelectric material (same acoustic impedance, same sound velocity) and identical density ρ. We obtain from the frequency $f(\mu)$

$$f(\mu) = \frac{v_s}{2}\left(\frac{1}{t_0 + \mu/\rho}\right), \; S_\mu = \frac{v_s}{2 f_0 \rho}\left(\frac{1}{t_p}\right)^2 = \frac{2}{v_s \rho} f_0. \qquad (5.15)$$

It is interesting to note that when using much more elaborate theories [14] the same result is obtained. Only the acoustic impedance of the resonator material matters. The relative sensitivity thus increases linearly with frequency, as obtained with the spring model for a constant $k_s m$. In discussing the detectivity or the SN ratio we must argue in a different way. We do not measure the mass loading directly; we observe it through its manifestation in the electrical signal. The part in the admittance (or impedance) that is due to acoustic interaction is proportional to the coupling factor k^2 [see (5.14) for the shear mode case]. Evaluation of the frequency shift can only be carried out to a precision that must be proportional to the width of the resonance peak (FWHM) amounting to f_0/Q. We can thus define a detectivity parameter D of the following form:

$$D = \frac{k^2}{\mathrm{FWHM}} \lim_{\mu \to 0} \frac{|\Delta f|}{\mu} \approx \frac{k^2 Q v_s}{2 f_0 \rho}\left(\frac{1}{t_p}\right)^2 = \frac{2 k^2 Q}{v_s \rho} f_0. \qquad (5.16)$$

Formally we still see the proportionality to the frequency. Unfortunately we have to consider that in any material, Q decreases with frequency as $Q = (2\pi f_0 \tau_1)^{-1}$, where τ_1 is a material's constant. It follows that

$$D \approx \frac{k^2}{\pi \tau_1 v_s \rho} \propto \frac{k^2}{\tau_1 c^{1/2} \rho^{1/2}} = \frac{k^2 c^{1/2}}{\eta \rho^{1/2}} = \frac{k^2}{\eta} v_s, \qquad (5.17)$$

where the viscous loss of the resonator material $\eta = c\,\tau_1$ [29] has been introduced. Such a detectivity is independent of frequency as long as the same materials and same sensor interrogation circuits are applied. The equation suggests that materials with a high sound velocity and small viscous losses are preferentially used. AlN is certainly one of the best possible choices. The additional materials in the resonator, above all the electrodes, also play an important role and are not included in this consideration. The exploitation of a larger sensitivity at higher frequencies works only if the electronics reading out the frequency shift works relatively more precisely at the higher frequency than at lower frequencies, meaning that smaller fractions of the FWHM peak width need to be detected.

Conclusions

Thin film bulk acoustic wave resonators based on AlN and ZnO thin films open new possibilities in gravimetric sensing. Strong miniaturization leads to much smaller sensor areas, enabling the fabrication of sensor arrays. Liquid delivery can be simplified by cointegration of fluidic channels in MEMS technology. As compared to the classical QCM, the miniaturization is accompanied by an increase in frequency of several orders of magnitude. However, this does not automatically lead to an increase in detectivity. The latter is rather a result of the materials in the resonator and of the performance of the electronics measuring the frequency shift. In addition, and not discussed in this article, one has to consider the impact of the immobilization layer on resonator performance.

References

1. Lu C-S (1975) Mass determination with piezoelectric quartz crystal resonators. J Vac Sci Technol 12:578–583
2. Pulker KH, Benes E, Hammer D, Söllner E (1976) Progress in monitoring thin film thickness with quartz crystal resonators. Thin Solid Films 32:27–33
3. Martin SJ, Ricco AJ, Niemzyk TM, Frye GC (1989) Characterization of SH acoustic plate mode liquid sensors. Sens Actuators 20:253
4. Vellekoop MJ (1998) Acoustic wave sensors and their technology. Ultrasonics 36:7–14
5. White RM, Wicher PJ, Wenzel SW, Zellers ET (1987) Plate-mode ultrasonic oscillator sensors. IEEE Trans UFFC 34:162
6. Gizeli E, Goddard NJ, Lowe CR, Stevenson AC (1992) A love plate biosensor utilizing a polymer layer. Sens Actuators B 6:131
7. Kovacs G, Venema A (1992) Theoretical cpmparison of sensitivities of acoustic shear wave modes for bio-chemical sensing in liquids. Appl Phys Lett 61:639
8. Wingqvist G, Bjurstrom J, Liljeholm L, Yantchev V, Katardjiev I (2007) Shear mode AlN thin film electro-acoustic resonant sensor operation in viscous media. Sens Actuators B 123: 466–473
9. Lakin KM, McCarron KT, Rose RE (1995) Solidly mounted resonators and filters. In: IEEE Ultrasonics Symposium. IEEE, Seattle, WA, USA
10. Gabl R, Feucht H-D, Zeininger H, Eckstein G, Schreiter M, Primig R, Pitzer D, Wersing W (2004) First results on label free detection of DNA and protein moecules using a novel integrated sensor technology based on gravimetric detection principles. Biosens Bioelectron 19:615–620
11. Weber J, Albers WM, Tuppurainen J, Link M, Gabl R, Wersing W, Schreiter M (2006) Shear mode FBAR as highly sensitive liquid biosensors. Sens Actuators A 128:84–88
12. Rey-Mermet S, Lanz R, Muralt P (2006) Bulk acoustic wave resonators operating at 8 GHz for gravimetric sensing of organic films. Sens Actuators B 114:681–686
13. Milyutin E, Gentil S, Muralt P (2008) Shear mode BAW resonator based on c-axis oriented AlN thin film. J Appl Phys 104:084508
14. Wenzel SW, White RM (1989) Analytic comparison of the sensitivities of bulk-wave, surface-wave, and flexural plate-wave utlrasonic gravimetric sensors. Appl Phys Lett 54:1976–1978
15. Kösslinger C, Drost S, Aberl F, Wolf H, Koch S, Woias P (1992) A quartz crystal biosensor for measurements in liquids. Biosens Bioelectron 7:397–404
16. Ballantine DS, White RM, Martin SJ, Ricco AJ, Zellers ET, Frye GC, Wohltjen H (eds) (1997) Acoustic wave sensors. Academic, San Diego
17. Harding GL, Du J (1997) Design and properties of quartz-based Love wave acoustic sensors incorporating silicon dioxide and PMMA guiding layers. Smart Mater Struct 6:716–720
18. Muralt P (2008) Piezoelectrics in micro and nanosytems: solutions for a wide range of applications. J Nanosci Technol 8:2560–2567
19. Muralt P (2000) Ferroelectric thin films for microsensors and actuators: a review. J Micromech Microeng 10:136–146
20. Drusedau TP, Blasing J (2000) Optical and structural properties of highly c-axis oriented nitride prepared by sputter deposition in pure nitride. Thin Solid Films 377:27–31

21. Dubois M-A, Muralt P (2001) Stress and piezoelectric properties of AlN thin films deposited onto metal electrodes by pulsed direct current reactive sputtering. J Appl Phys 89:6389–6395
22. Takikawa H, Kimura K, Miyano R, Sakakibara T, Bendavid A, martin PJ, Matsumuro A, Tsutsumi K (2001) Effect of subsgtrate bias on AlN thin film prepariation in shielded reactive vacuum arc deposition. Thin Solid Films 386:276–280
23. Artieda A, Barbieri M, Sandu CS, Muralt, P (2009) Effect of substrate roughness on c-oriented AlN thin films. Journal of Applied Physics, 105, 024504.
24. Bjurstrom J, Wingqvist G, Katardjiev I (2006) Synthesis of textured thin film piezoelectric AlN films with a nonzero c-axis mean tilt for the fabrication of shear mode resonators. IEEE Trans UFFC 53:2095–2100
25. Yanagitani T, Kiuchi M, Matsukawa M, Watanabe Y (2007) Shear mode electromechanical coupling coefficient k15 and crystallites alignment of (11–20) texture ZnO films. J Appl Phys 102:024110
26. Bjurstrom J, Rosen D, Katardjiev I, Yanchev VM, Petrov I (2004) Dependence of the electromechanical coupling on the degree of orientation of c-texture thin AlN films. IEEE Trans UFFC 51:1347–1353
27. Aigner R, Kaitila J, Ellia J, Elbrecht L, Nessler W, Handmann M, Herzog T, Marksteiner S (2003) Bulk-acoustic wave filters: performance optimization and volume manufacturing. IEEE IMS 3:2001–2004
28. Benetti M, Cannata D, DiPietrantionio F, Foglietti V (2005) Microbalance chemical sensor based on thin-film bulk acoustic wave resonators. Appl Phys Lett 87:173504
29. Ballato A, Gualtieri JG (1994) Advances in high-Q piezoelectric resonator materials and devices. IEEE Trans UFFC 41:834–844

Chapter 6
Lab-on-a-Chip for Analysis and Diagnostics: Application to Multiplexed Detection of Antibiotics in Milk

Janko Auerswald, Stefan Berchtold, Jean-Marc Diserens, Martin A.M. Gijs, Young-Hyun Jin, Helmut F. Knapp, Yves Leterrier, Jan-Anders E. Månson, Guillaume Suárez, and Guy Voirin

Introduction

The concern for "food safety" surely emerged very early in human history, contributing to the establishment of certain rules and customs. In today's technologically developed countries, food safety is subject to strict laws that regulate the presence of undesired substances in food. In particular, in the milk industry, levels of residues of veterinary medicinal products, of which antibiotics represent a significant part, are regulated by European Council (EC) Regulation no. 2377/90. More precisely, a series of four antibiotic families are found to be of particular interest due to their routine use for treatment in bacterial infection and/or prophylactic purposes: fluoroquinolones, sulfonamides, β-lactams, and tetracyclines. Their excessive use in dairy diet in recent decades gave rise to stronger bacterial resistance, which consequently represents a serious problem in the efficiency of classic antibacterial treatment in humans [1]. In this context, there is a considerable need for developing sensing devices able to detect a series of antibiotic families simultaneously. A few dipstick format tests [1] commercially available that are cheap, fast, and reliable currently offer an interesting alternative to expensive conventional chromatographic techniques. However, those systems are not fully automated and are unable to detect more than two antibiotic families per single test. The present work reports the development of a lab-on-a-chip (LOC) test system for multiplexed detection of four antibiotic families in raw milk. A fabricated microfluidic cartridge is prefilled with the solutions necessary for immunoassays, and the whole detection sequence is operated automatically by external pump-and-valve combinations. The immunoassay principle is based on a competitive assay format, and the resulting refractive index changes at the sensor surface are probed using the wavelength interrogated optical sensing (WIOS) method. The multiplexed sensing system that was consequently developed was adapted for the simultaneous detection of sulfonamides, fluoroquinolones, β-lactams, and tetracyclines. The whole test procedure is

Guy Voirin (✉)
CSEM – Swiss Center for Electronics and Microtechnology Inc., Neuchâtel, Switzerland

G. De Micheli et al., *Nanosystems Design and Technology*,
DOI 10.1007/978-1-4419-0255-9_6, © Springer Science+Business Media LLC 2009

fast (less than 10 min), easy to handle (automated actuation), and cost-effective due to the use of novel photosensitive materials in a "one-step" fabrication process.

The Problem of Antibiotics in Milk

The problem of antibiotics in milk was recognized in the 1950s, while the first systematic test of antibiotic in raw milk started in the 1960s. The problem of having antibiotics in milk has at least three facets:

- It can cause allergic reaction in sensitized people; about 8% of patients trigger an allergic reaction after the first administration of β-lactam, a major class of antibiotics. Among them, 15 to 20% will present a severe reaction at a new administration. Thus it is important for such people to avoid any exposure to β-lactam.
- More and more pathogens are becoming resistant to antibiotics. Antibiotic residues are suspected to be at the root of this resistance. In fact, people have helpful bacteria in their intestines and antibiotic residue can favor the survival of antibiotic-resistant strains that can be transferred to harmful bacteria, causing some people to develop illnesses [1].
- A low level of antibiotics in milk can partially or totally inhibit the fermentation of leaven used for the manufacture of cheese or yogurt. This can be the cause of important commercial losses for the dairies concerned.

Antibiotic residues have found their way into milk from a variety of sources. Antibiotics are used by farmers to maintain the health of their livestock. Milking cows can suffer from mastitis, and the injection of antibiotics directly into the udder of such cows is often necessary. The milk from such cows will then be contaminated for several days. For some other illnesses, antibiotic injections are sometimes necessary. These injections are made in the muscles and their path to the milk is longer and the contamination will therefore be shorter. Other habits among farmers are the preventive treatment of cows during the time when they are susceptible to contracting mastitis. Sometimes antibiotics are also preventively added to cattle feed to improve production.

In order to meet current regulations, authorities have implemented a control policy that changes from country to country. Generally each milk producer is checked several times per month for antibiotic residue in delivered milk. At the dairy entrance, each truck is checked before unloading. In there is contamination, its origin will be traced and the farmer will be severely penalized (a small milk sample of each farmer is set aside during collection). If it is suspected that the contamination arose due to problems during milking, poor identification of treated cows, etc., the farmer will be given the opportunity to perform tests at the farm before delivering milk from his or her cows.

Milk screening at the farm level is linked to the frequency of positive cases and to the penalty imposed on the farmer when he is pronounced guilty. Systematic testing at the farm level will only be affordable if the price of the test is low and if it is in a country where the rate of positive cases is high. The time saving obtained by testing milk in trucks or at farms is currently not so great and it will be advantageous only if a range of antibiotics can be tested.

In the framework of the European GoodFood project, a survey was conducted by Nestlé using a questionnaire that was sent to major reference laboratories and world-wide companies active in the field of milk. The main results concern the importance of several antibiotic families: β-lactam, tetracycline, sulphonamide, quinolone, and amphenicol (macrolides and aminoglycosides are often used in combination with one or the other). The measurement must be made in 5 to 10 min and the detection limits of interest to users are the maximum residue limits (MRL), which are the legal limits, or the minimum required performance limits (MRPL), which are the minimum concentration to be detected to have a method considered as sufficiently sensitive.

To be able to comply with the regulations, reference laboratories have developed reference methods, mainly based on high performance liquid chromatography (HPLC) or mass spectrometry (MS). Numerous antibiotic-detection protocols have been developed or are undergoing continual improvement based on these conventional techniques, such as HPLC coupled with UV spectroscopy (HPLC–UV) [2] or MS (HPLC–MS) [3]. Despite their high performance and their ability to identify a large number of compounds, these techniques usually require a minimum of sample preparation, must be run by trained technicians, and must be used in a laboratory environment.

Another group of standard test methods used for antimicrobial contaminant detection in milk for official quality control is based on microbial inhibition assay. Delvotest is among those commercially available tests that offer broad-spectrum screening covering all major groups of antibiotic families [4]. This type of test suffers from a long incubation time (several hours), which is not affordable in the milk production chain. A truck cannot be stopped so long before unloading.

A last class of antibiotic residue detection system is based on immunoassay. The four main challenges for these devices are:

- Antibodies or receptor elements must be selected that are sensitive to not just one antibiotic but a whole family of antibiotics.
- Parallelization: indeed as these types of devices are based on the selective binding of engineered or natural molecules with antibiotics, it is inconceivable to find a molecule able to selectively bind to all classes of antibiotics. Therefore it will be necessary to develop multisensor and select reagents to avoid cross reactivity;
- The time to result must be shorter than 10 min.
- They must be easy to use.

Recently immunoassay tests for the detection of multiple antibiotics have been developed based on enzyme-linked immunosorbent assay (ELISA) [5] and parallel affinity sensor array (PASA) [6]. Despite their reliability, sensitivity, and capacity to select for more than ten different antibiotics, those tests must be run in a laboratory

Table 6.1 Commercialized rapid test kit (adapted from Food Safety Authority of Ireland, 2002)

Product	No. of antibiotics	Test time (min)	Cost per test	Quant.
Parallux (Medexx)	1(2, 3)	4	€4 (€4.4, €5.7)	
SNAP (IDEXX Inc.)	1	10	€3.6	No
Charm MRL (Charm Science)	1	9	€3.5–7.72 (depends on antibiotic)	No
Beta S.T.A.R (NEOGEN)	1	5	€2–3.4 (depends on volume of order)	No
TwinsensorBT (UNISENSOR)	2	10	€3.5	No
GOAL	>4	<10	€3–5/antibiotic family	Semi

environment, necessitate a large number of manipulations, and require at least 1 h for assay time.

To meet the milk industry market demand for rapid and cheap detection devices, several tests based on lateral flow immunochromatography have appeared on the market. The main advantage of those strip tests or "dipsticks" is their ease of use. The milk sample is mixed with immunoreagents (labeled with colloidal gold or carbon) and capillary-driven through a nitrocellulose strip. During the diffusion, the reagents interact with the milk and, depending on the interaction, will be trapped or not on a functionalized region of the strip. The response of the test can be read visually by comparison with a reference region. A specific reader can also be used to obtain an electronic reading without human judgment. Typically, those binary response tests are fast (less than 10 min), easy to perform, cheap, sensitive, and specific. Several years ago, a rapid test in dipstick format was commercialized for the simultaneous detection of two families of antibiotics (β-lactams and tetracyclines) [7]. However no commercial system is able to meet all the challenges simultaneously: detecting a high number of antibiotic families, ease of use, and short time to produce results (Table 6.1).

The work reported here presents a device that fits in this context and that meets the challenges. It is based on:

- A LOC device for ease of use and minimum manipulation by the user
- An optical transduction technique for direct reading of receptor target reaction (without the need for a development step as when using enzymatic reaction) for quick measurement
- Antibodies and receptors specially engineered to be selective to a large number of congeners inside an antibiotic family to be able to detect a large number of antibiotics with a reasonable number of different measurement regions

The whole immunoassay protocol is carried out automatically in a microfluidic cartridge that is operated by external pump-and-valve automated actuation. Reservoirs for reagent solutions and a sensor chip for optical detection are integrated in the microfluidic cartridge. The optical transduction system makes use of the recently developed WIOS technology [8,9]. The detection of antibiotics with the WIOS system was demonstrated using an immunoassay format for sulfonamides [10]. The reagents have already been used in a similar format either in ELISA or with dipstick, and four antibiotic families are covered: sulfonamide, quinolone, β-lactams, and tetracycline. Unlike conventional techniques such as HPLC, microbial inhibition assay, ELISA, and PASA, the developed system offers an easy-to-use and fast test allowing detection in a single operation of a large number of antibiotics from four different families at a reasonable cost compared to the price of a single test.

Detection

Optical Transduction Principle

Optical detection is based on the recently developed WIOS approach [9]. The measurement takes place at the interface between a waveguide and a liquid, if a layer of molecules is created at this interface; optically this is equivalent to a thin layer with a refractive index specific to this molecule generated at the liquid waveguide interface. This will affect the propagation properties of the waveguide mode and in particular the propagation velocity. The sensing principle is twofold. First a biomolecular interaction that will depend on the antibiotic contamination of the sample must take place at the interface. The result of this interaction will be a change in the thickness and/or the refractive index of the thin layer at the liquid waveguide interface. The specificity and sensitivity of the sensor system are given by the quality of the sensing layer. The sensing layer can be composed of recognition molecules like binding proteins, antibodies, DNA strands, or chemical receptors. Second, a grating coupler will be used to sense the variation in the waveguide mode velocity. Different optical configurations using waveguide grating couplers have been studied, looking at the coupling/outcoupling angle to the waveguide [11] or using a chirped grating and looking at the coupling position (light pointer) [12]. In this work, the developed system monitors the wavelength (resonance wavelength) that is coupled to a waveguide grating [8,9].

   If the resonance condition is fulfilled, light is transmitted via an output grating coupler to a photodetector. Due to the continuous wavelength scanning, the output signal of the photodetector represents the grating coupler resonance peak whose position is used to estimate the propagation constant of the waveguide mode. Changes in the peak position indicate variation in the quantity (surface mass density) of analyte molecules adsorbed on the sensor chip's surface. The peak position corresponding to the resonance wavelength for the coupling to the input waveguide grating is governed by the following grating equation:

$$\lambda_r(t) = \Lambda(n_e(t) - \sin\theta), \qquad (6.1)$$

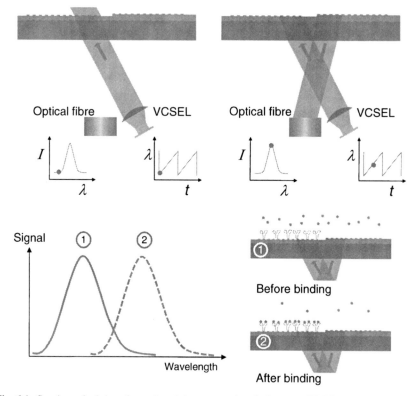

Fig. 6.1 Sensing principles of wavelength interrogated optical sensor (WIOS) technology

with λ_r the resonance wavelength for which coupling occurs, Λ the grating period, θ the incidence angle, and $n_e(t)$ the effective index of the waveguide mode ($\frac{2\pi}{\lambda} n_e$ propagation constant of the waveguide mode).

A small system was built with a vertical cavity surface emitting laser (VCSEL) as wavelength tunable laser source, a plastic optical fiber for collecting the outcoupled light and guiding it to photodetectors, an electronic circuit for driving the laser current with a periodical sawtooth function, and a data acquisition card in a computer for processing the photodiode signals. The peak position is extracted and displayed as a function of time on the computer screen. During the measurement, the optical configuration is fixed (θ, Λ fixed) and the monitoring of λ_r gives access to effective refractive index variations of the waveguide mode due to the molecular binding occurring at the waveguide-bulk interface, as shown in Fig. 6.1.

The waveguide grating chips were purchased from Unaxis Optics (now Oerlikon OC). The grating structures are fabricated using holographic lithography followed by a dry etching technique on the AF45 glass substrate [9]. The grating structure is then coated with a high refractive index layer (Ta_2O_5 dielectric film) to form the waveguide layer. Each sensing chip ($17.5 \times 17.5 \, mm^2$) is constituted of eight individual sensing regions (grating pads) allowing simultaneous and real-time monitoring of up to eight different molecular interactions.

Immunoassay Formats

Individual and multiplexed competitive immunoassay formats for antibiotic detection in milk based on a microplate assay approach [5] and on WIOS technology [13] have already been reported in the literature. Three of the assays were performed in indirect formats where specific haptens for sulfonamides, fluoroquinolones, and β-lactams were coated onto different sensing regions (gratings) of the chip. In the case of tetracyclines, a specific DNA sequence was attached to a sensing region via biotin–neutravidin interaction. The chip biofunctionalization was carried out using a polysaccharide-based photolinker polymer (OptoDexTM, Arrayon Biotechnology) thin-film coating followed by a nanoplotting technique.

Despite the use of different capture-analyte configurations, all the immunoassays developed were based on a competitive format. For sulfonamide and fluoroquinolone detection, the antibiotic in solution competes with the hapten immobilized on the surface to bind to a specifically designed antibody added to the sample mixture, as depicted in Fig. 6.2a. Specific receptors are used for the detection of β-lactam and tetracycline antibiotics. In the case of β-lactams, the competition occurs between an immobilized hapten-antigen and the antibiotic in solution to bind specifically to a receptor, as shown in Fig. 6.2b. Finally, for tetracyline detection,

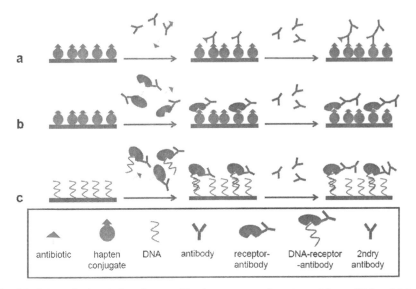

Fig. 6.2 Schematic description of competitive immunoassay formats used for multiplexed detection of sulfonamides (**a**), fluoroquinolones (**a**), β-lactams (**b**), and tetracyclines (**c**)

the antibiotics in solution compete with DNA strands to link specifically to a DNA-modified receptor. However, the receptor–antibiotic complex can no longer bind to the DNA surface and is subsequently washed off during the assay (Fig. 6.2c). Prior to introduction into the fluidic system the milk to be analyzed was mixed (1:5 dilution) with a "cocktail" solution containing the different antiserum receptors that specifically react with their corresponding antibiotics, if any are present in the sample. After a washing step the attached antibodies were revealed by reacting with a secondary antibody. Typically, as for any competitive assay, the higher the concentration of analyte (a given antibiotic), the lower the signal obtained, reflecting a low amount of free antibody able to attach to the sensing surface.

Microfluidic Cartridge Fabrication

State of the Art

Considerable efforts have been undertaken in recent years to substitute silicon or glass by polymer materials for the fabrication of LOC devices. A broad variety of polymers have been explored for their suitability in the manufacture of LOC, including polyimide [14], polymethylmetacrylate (PMMA) [15–21], polydimethylsiloxane (PDMS) [22–24], polyethylene [25, 26], PC [27, 28], or cyclic olefin copolymer (COC) [29, 30]. One of the significant advantages of using polymers is the low fabrication cost through replication approaches. However, compared to silicon and glass, polymer LOC devices often lack dimensional stability due to uncontrolled process-induced internal stresses resulting from differential thermal contraction and cure shrinkage. UV-curing processes, which are generally carried out at room temperature, solve the thermal contraction problem. However, the usual negative-tone photoresists (epoxy based SU-8 and Novolak) still suffer from very high shrinkage stress, which compromises the dimensional accuracy of produced microstructures. This may be overcome through long postbaking cycles, at the expense of cost-effectiveness. An alternative was recently introduced in the form of UV-curable hyperbranched polymers (HBP). These materials combine fast conversion and low shrinkage owing to delayed gelation and vitrification during network formation, which proved to be very effective for producing a variety of device structures with high dimensional accuracy [31–34].

Dendritic Hyperbranched Polymers

HBPs are a class of dendritic macromolecules, characterized by a regularly branched treelike molecular structure [35]. Since the pioneering efforts in the early 1980s, worldwide research on dendrimers has developed to a considerable extent.

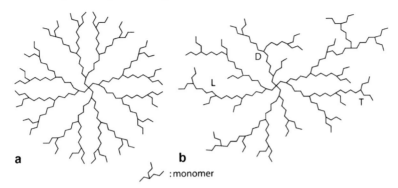

Fig. 6.3 Schematic structures of a second-generation dendrimer (**a**) and hyperbranched polymer (**b**). One dendritic, one linear, and one terminal unit D, L, T are indicated. Adapted from Rodlert et al. [36]

Dendrimers have a perfect structure with a degree of branching and polydispersity of one (Fig. 6.3a), whereas HBPs (Fig. 6.3b) have less perfect structures with a degree of branching less than one. In spite of their imperfect structure, HBPs are more cost-effective owing to their considerably easier synthesis compared to dendrimers.

The viscosity of dendrimers and HBPs is lower than that of their linear counterparts [37], and they behave more like Newtonian fluids due to fewer entanglements [38, 39]. These properties are particularly useful for producing low-cost micro-/nanodevices using replication methods.

Micro-/Nanostructuring of Hyperbranched Polymers

The fabrication methods for micro-/nanostructures using polymer materials can be categorized as direct patterning methods and replication methods. In the direct patterning method, the polymer material is selectively cured to form a micro-/nanostructure using UV light (lithography) [40–43], ion beam etching [44], and laser (stereo lithography) [45, 46]. Replication methods involve a so-called master, which has inverse micro-/nanopatterns of the final structure. The fabrication of the master is a costly process, but it can be used many times repeatedly and thus has good potential for mass production of cheap disposable devices. The most popular replication methods include casting [22–24], nanoimprint lithography (NIL) [47], and injection molding [30].

Several HBP materials have been selected in recent years to manufacture microdevices. Hyperbranched perfluoropolyether (PFPE) was used to produce microfluidic devices [48, 49]. The PFPE material is soft but resistant to organic solvents and was found to be suitable for soft lithography applications. The HBP used for the LOC device discussed in this chapter was investigated in previous works [50, 51]. In addition to negligible thermal shrinkage, this material exhibits

Table 6.2 Physical properties of acrylated polyether HBP

Property	Unit	
Theoretical functionality		32
Actual functionality		29
Mn	g/mol	3'577
Mw	g/mol	8'521
PDI		2.4
Newtonian viscosity	Pa s	6
AEW	g/mol	294
DB		0.4
T_g (monomer)	°C	−55

Mn number molecular weight; *Mw* mass molecular weight; *PDI* polydispersity; *AEW* acrylate equivalent weight; *DB* degree of branching [52]; T_g glass transition temperature. Adapted from [53]

low network formation shrinkage, resulting in low deformation and distortion after fabrication. A further potential of the HBP is its hydrophilic nature. It is based on a third-generation hyperbranched polyether polyol that was synthesized and acrylated by Perstorp AB, Sweden, giving a 29-functional UV-curable polyether acrylate. Table 6.2 gives an overview of the physical and chemical properties of the HBP. The number of acrylate functions per monomer and the molecular weight were in accordance with the specifications of the supplier. A photoinitiator was selected to match the absorption spectrum of the HBP (Irgacure 500, a mixture of equal parts of 1-hydroxy-cyclohexyl-phenyl-ketone and benzophenone, Ciba Specialty Chemicals, Switzerland), at a concentration equal to 2 wt%. It was blended with the acrylate monomer at a temperature of 85°C to facilitate mixing.

Microstructuring of the acrylated HBP was found to be feasible using both replication methods (micromolding or casting) [53] and direct patterning [54]. In the micromolding process [51] (Fig. 6.4), a master was fabricated by an SU-8 lithography process in a first step. The liquid HBP monomer was subsequently dispensed on the master and exposed to UV light. The thickness of the monomer layer was controlled by spacers, resulting in 33-μm-thick HBP microstructures, with 14.5-μm-wide vertical walls, 14.7-μm-wide fluidic channels, 24.1-μm-wide square pillars, and 53.4-μm-wide square holes. A microfluidic network device, composed of microfluidic channels and reservoirs, was also fabricated and its microfluidic performance was verified by a fluidic test.

The acrylated HBP was also evaluated to produce ultrathick microstructures using a lithography process [54]. The objective was to evaluate the potential of the HBP as photoresist for thick polymer microstructures with a reduced internal stress, with attention paid to dimensional accuracy and processing time. In the first step of fabrication, to improve adhesion between the polymer and the glass or silicon substrate, a thin layer of hexamethyldisilazane and a 30-μm-thick layer of HBP were deposited on the substrate. The microstructures were produced after exposure through a mask comprising the micropatterns. Glass spacers were used to control the thickness of the liquid monomer and a 12-μm-thick polyethyleneterephthalate

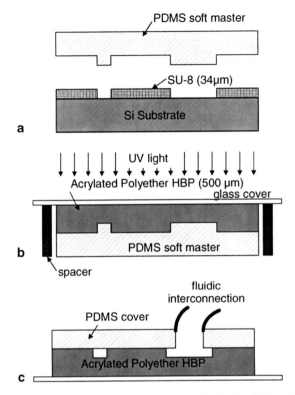

Fig. 6.4 Fabrication process of fluidic devices using acrylated polyether HBP and a PDMS master. **a** PDMS molding for a soft master. **b** UV curing of acrylated polyether HBP. **c** Bonding with a PDMS cover for the fluidic interconnections. Adapted from Jin et al. [53]

film was placed between the mask and monomer to protect the mask (Fig. 6.5). After exposure, the device was placed in the development solution (1-methoxy-2-propyl acetate). Microstructures with thicknesses of up to 850 μm, aspect ratios of up to 7.7, a 5.5-fold reduction in internal stress, and a sixfold reduction in processing time compared to SU-8 were demonstrated. Moreover, a high-performance valve for microbattery devices was produced using this UV-curable HBP.

Hybrid HBP/PMMA Microfluidic Cartridge

The microfluidic cartridge is the core of any LOC system since it is the place where the chemical or physical operation(s) will take place. However, growing developments in fabrication techniques and the large variety of applications have enlarged considerably the possibilities of material, design, or integrated functions for a microfluidic cartridge fabrication. In the present context the microfluidic

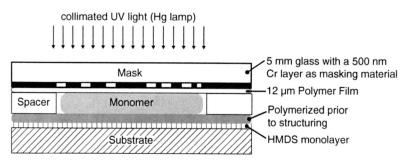

Fig. 6.5 Photolithographic fabrication method. The collimated UV light was produced from an Hg lamp with an intensity maximum at 365 nm (Hg i-line). Adapted from Schmidt et al. [54]

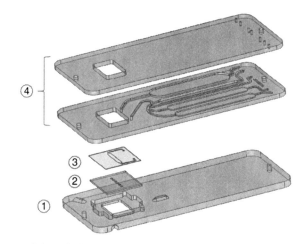

Fig. 6.6 Fragmented view of microfluidic cartridge showing constitutive elements: (1) holder plate for sealing and chip positioning; (2) WIOS chip; (3) gasket defining sensing chamber on WIOS chip; (4) reservoirs plate containing reservoirs, channels, and inlet/outlet flat ports

cartridge has to contain reservoirs filled with the different reagents required for the immunoassay, a sensing chamber giving access to the sensing areas, waste reservoirs to restrain contamination, and a channel network that defines flow pathways. The liquid motion is externally actuated by pressure control, keeping clear the use of integrated microvalve/pump that would have increased the cost of the device. As depicted in fragmented view in Fig. 6.6, the microfluidic cartridge developed consists of the assembly of several elements. The top element is a reservoir plate containing reservoirs, channels, and inlet/outlet flat ports. The intermediate element is a gasket connected to the reservoir plate that defines the sensing chamber on a WIOS chip. On the bottom is a holder plate that has no fluidic features to ensure sealing and chip positioning on the instrument. For the reservoir plate, which exhibits microstructures in the range of several hundred micrometers to 1 mm, the choice of material led to PMMA. The reservoirs and delivery channels were shaped by micromilling and sealed by thermal bonding (Fig. 6.7a). The external dimensions

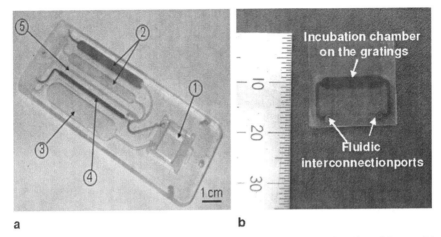

Fig. 6.7 Photograph of lab-on-a-chip (LOC) cartridge for simultaneous detection of three antibiotics in raw milk. **a** Micromilled PMMA cartridge combined with sensor chip: (1) sensor chip (optical transducer with incubation chamber), (2) reservoirs for immunoassay solutions, (3) main waste, (4) loading waste, (5) sample introduction channel. **b** Enlarged view of sensor chip (HBP incubation chamber bonded on the grating chip)

of the cartridge are $100 \times 40 \times 7 \, mm^3$ and reservoirs exhibit a volume capacity of around $100 \, \mu L$. Channels have a rectangular-shaped section with dimensions of $0.9 \times 0.5 \, mm^2$.

In the sensing chamber of the sensor chip, laminar flow conditions of the solutions should prevail to ensure stable and reproducible measurements. Thus, the sensing chamber was made using the UV-curable HBP described in the previous section, owing to its outstanding dimensional stability. A novel fabrication technique combining UV micromolding and photolithography was developed for the simultaneous fabrication of microchannels and interconnection ports (Fig. 6.7b). This novel technique combines the advantage of the conventional micromolding process [53] and lithography process [54], i.e., easy pattern transfer for microchannels and via hole making, without the drawbacks of expensive and time-consuming backend processes such as drilling. The micromolding master was fabricated by lithography with a 100-μm-thick SU-8 layer (Gersteltec, Switzerland). The liquid HBP monomer was dispensed on the SU-8 master and exposed for $20 \, s$ to UV light with an intensity of $50 \, mW/cm^2$ through a Cr photomask with the patterns of the interconnection port (Fig. 6.8b-1). Thereafter, the photomask was removed and the uncured part of the HBP was developed using propylene glycol monomethyl ether acetate (PGMEA, Aldrich Chemical Co., Milwaukee, WI, USA) solution (Fig. 6.8b-2). In the final step, the sensing chamber was generated on the biofunctionalized sensing chip by bonding the HBP microchannel using UV adhesive (Dymax, Torrington, CT, USA). The sensor chip was assembled in the reservoir plate (Fig. 6.8c) and the reservoirs filled with the different solutions required for the immunoassay.

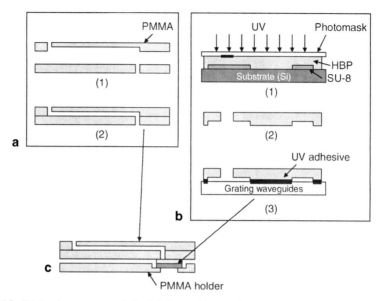

Fig. 6.8 Fabrication process of the LOC cartridge. **a** Reservoir plate: (1) micromilling of PMMA plate for reservoirs and microchannels, (2) thermal bonding for sealing of reservoirs and microchannels. **b** Sensor chip with sensing chamber: (1) UV curing of HBP on SU-8 through photomask for sensing chamber, (2) development of uncured HBP, (3) bonding of sensing chamber on grating waveguides chip using UV adhesive. **c** Assembly of reservoir plate and sensor chip

Nanosized HBP Grating Waveguide

The sensor chip is composed of grating waveguides with a 360-nm period produced on a glass substrate. The grating patterns are defined by holographic exposure and transferred to the glass substrate by dry etching [9]. The glass grating shows very good mechanical and chemical stability, which is essential for highly sensitive sensor applications. However, in terms of fabrication cost, the glass grating chip is not compatible with disposable devices.

Because of their low viscosity and low polymerization shrinkage, HBPs are good candidates for the fabrication of nanosized gratings with comparable dimensional stability. Their very fast conversion into a solid should also be extremely advantageous. In this context, the feasibility of the HBP replication process to produce the grating structure was evaluated. The acrylated HBP material was prepared as described in the previous section and the glass sensor chip was used as a master. The HBP grating was replicated from the glass master by the NIL method using a low-pressure, photocuring imprint machine. Figure 6.9 compares the glass master grating with an example of a replicated HBP grating obtained after 20 s exposure to a UV intensity of $30\,\mathrm{mW/cm^2}$ under 3 bar. In fact, high-intensity UV lamps such as those used in industrial processes would enable one to reduce the process time to less than 1s. An excellent dimensional fidelity is evident, which demonstrates

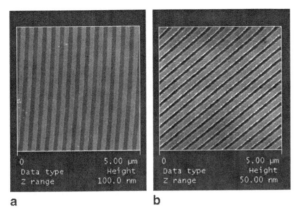

Fig. 6.9 Atomic force microscope topographic images of a glass grating fabricated by holographic exposure and dry etching technique (**a**) and a HBP grating fabricated by fast UV nanoimprint replication technique (**b**)

the relevance of this new class of polymer materials for very fast and cost-effective fabrication of nanoscale structures.

Fluidic Setup

Flow Motion Actuation

Fluidic setup embraces a large domain dedicated to the development of systems able to provide a controlled flow motion along the microfluidic cartridge channel pathways. Among the numerous approaches developed for driving inner flows, the on-chip electro-osmotic pump [55, 56] is surely one of the most elegant. These integrated pumps are particularly convenient for low-flow-rate devices (less than $1\,\mu L\,min^{-1}$) but generally require a high voltage source (up to 1 kV) and the fabrication process tends to increase the cost of the microfluidic chip. In this context, pressure-driven fluidic setups represent an interesting alternative, particularly suitable for microfluidic cartridges with a "passive" layout. This approach allows easy control of the flow rate and avoids using costly active channels on a chip. Pressure-driven fluidic setups have been abundantly reported in the literature and for many applications. Even in the narrow scope of this study, such a system has been developed for a microarray system actuated by eight syringe pumps for automated analysis of multiple antibiotics in milk [6]. The microfluidic cartridge described in the previous section contains reservoirs filled with reagent solutions that have to move in a sequential way to the sensing chamber where multiplexed immunoassay takes place. For this purpose, a pressure-driven fluidic setup has been developed where fluid motion is driven by a single syringe pump (CavroTM XCalibur) working in aspiration mode in combination with a multi-position valve (CavroTM Smart Valve) that selectively connects the reservoirs to the atmosphere, as depicted in Fig. 6.10. Both pump and valve are controlled

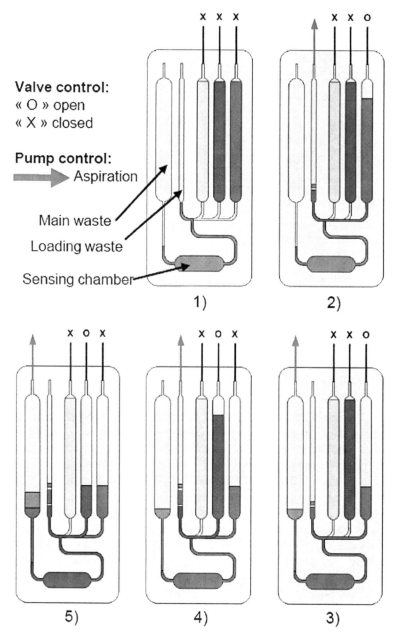

Fig. 6.10 Schematic representation of fluidic setup for sequential liquid delivery from prefilled reservoirs to sensing chamber. The configuration of microfluidic cartridge during storage is depicted in (1) with sensing chamber filled with buffer solution. During experiment the whole sequence is controlled by external and concerted actuation of one multiposition valve connected to atmospheric pressure and one syringe pump (three ports). Air plug is deviated toward loading waste for continuous liquid–liquid junction on sensing region

through PumpLink32 software (CavroTM) that allows programming a sequence of coordinated operations.

The introduction of a "loading waste" in the channel network layout allows for easy removal of air plugs between reservoirs. Prior to going to the sensing chamber for reaction/measurement the flow is deviated toward the loading waste until no residual air remains in the general channel path. The efficiency of this simple "diverted flow" approach was experimentally demonstrated by following visually the motion of colored solutions through the cartridge, as shown in Fig. 6.11. This fluidic setup allows the continuous delivery of the sample volume through the sensing chamber with no air plug between adjacent solutions.

Moreover, waste reservoirs ensure that no residual liquid leaves the cartridge, thereby limiting the risk of contamination. Fluidic connections between the cartridge and the valve/pump system are ensured through a so-called "holder interface" located on the WIOS instrument, as shown in Fig. 6.12. Once the cartridge is positioned in the fluidic holder box the cover is closed by applying some pressure on the rubber O-rings of the "fluidic adaptor" piece; these O-rings seal the connection between the inlet/outlet flat ports of the cartridge and external tubing.

Under laminar flow conditions, the reaction rate depends principally on the analyte concentration, leading to a higher stability and reproducibility of the measurement. Regarding the present system, the hydrodynamic flow generated by negative pressure actuation through the microfluidic channels and incubation chamber is laminar (flow rate set at $25\ \mu L\ min^{-1}$), as indicated by the very low Reynolds number, $R_e = 0.85$, calculated from the following equation:

$$R_e = \frac{4Q}{\pi d_h v}, \tag{6.2}$$

where Q represents the volumetric flow and d_h and v are the hydraulic diameter and the dynamic viscosity, respectively.

Sample Introduction

A small volume of milk sample ($40\ \mu L$) is introduced by pipetting into a sample vial ($250\ \mu L$), which is previously filled with the assay reagents required for competitive immunoassay. The sample mixture is then shaken and the septum cap of the vial is plugged on the two needle inlets of the "fluidic adaptor." One of the needles is connected to the multiposition valve while the other reaches the sample channel of the microfluidic cartridge, as depicted in Fig. 6.13. In this way, the sample is easily connected to the fluidic system and can be delivered to the LOC through external pump/valve actuation.

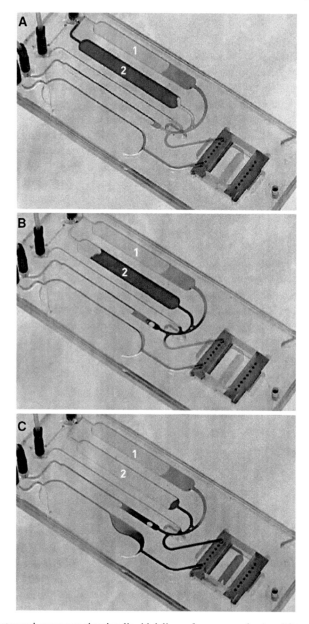

Fig. 6.11 Photograph sequence showing liquid delivery from reservoirs 1 and 2 to sensing chamber. **a** Reservoir 1 connected to air: solution successively delivered to loading waste and sensing chamber. **b** Reservoir 2 connected to air: solution 2 driven to loading waste. No residual air between solutions 1 and 2. **c** Solution 2 driven to sensing chamber (Fig. 6.7)

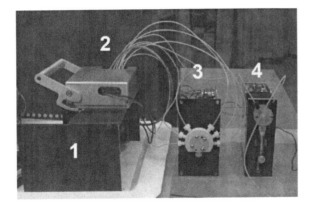

Fig. 6.12 Photograph of automated multidetection system: (1) WIOS detection instrument; (2) cartridge inserted into "holder interface" box; fluidic connections between microfluidic cartridge and "holder interface" are ensured by rubber O-rings of "fluidic adaptor"; (3) multiposition valve; (4) syringe pump

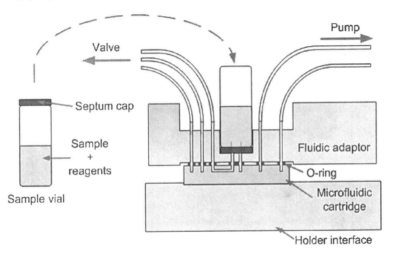

Fig. 6.13 Schematic representation of milk sample introduction into fluidic setup

Measurements

Test Protocol

The full protocol for LOC detection of multiple antibiotics in milk is summarized in a limited number of steps in Fig. 6.14. The prefilled microfluidic cartridge is first inserted and positioned onto the "holder interface," which is subsequently closed. A fixed volume (40 μL) of raw milk sample is transferred into a vial that contains the competitive reagents for immunoassays. Afterwards, the vial is plugged through

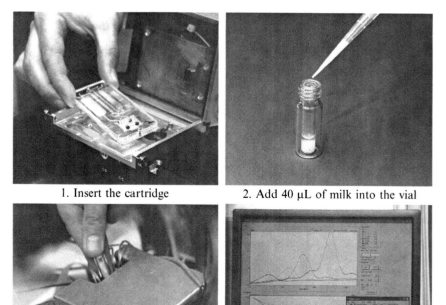

| 1. Insert the cartridge | 2. Add 40 μL of milk into the vial |

| 3. Plug the vial on the instrument | 4. Run program and read results |

Fig. 6.14 Sequence of pictures illustrating test protocol for LOC multiplexed detection of antibiotics in milk

its septum cap into the specific location of the "fluidic adaptor." Finally, the laser source of the WIOS instrument is turned on and the program for concerted pump-and-valve actuation is initiated. The whole automated multiplexed immunoassay takes place within 10 min.

Test Interpretation

The whole automated immunoassay sequence depicted in Fig. 6.15 consists of the following steps: baseline stabilization (buffer), milk sample introduction, washing, and introduction of secondary antibody for signal amplification. The differential signal value is measured at 10 min from program initialization. Typically for a binary response test the value obtained at maximum residue limit (MRL) concentration defines two qualitative domains: for any value less than or equal to the value at MRL the test is considered "positive," the remaining domain being "negative."

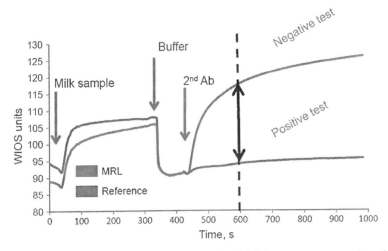

Fig. 6.15 Typical WIOS response curve obtained with LOC immunoassay test where threshold curve obtained at maximum residue level (MRL) concentration defines positive and negative domains

Table 6.3 WIOS differential responses calculated from response curves of Fig. 6.16

	Sulfapyridine		Ciprofloxacin		Ampicillin		Oxytetracycline	
0 ppb	19.9[a]	100[b]	44.4 [a]	100[b]	10.6[a]	100[b]	8[a]	100[b]
4 ppb	–	–	–	–	7.8[a]	75[b]	–	–
100 ppb	6.5[a]	32[b]	11.4[a]	26[b]	–	–	3[a]	37[b]

[a] Absolute value
[b] Normalized value

Multiplexed Detection Results

The reliability of the LOC multiplexed detection system was demonstrated for four antibiotic families: sulfapyridine (sulfonamide), ciprofloxacin (fluoroquinolone), ampicillin (β-lactam), and oxytetracycline (tetracycline). For the four integrated biodetection systems the signal obtained in the absence of antibiotics in milk (maximum response) was compared with values obtained with milk spiked at MRL concentrations (threshold response): 100, 100, 4, and 100 ppb for sulfapyridine, ciprofloxacin, ampicillin, and oxytetracycline, respectively. The response curves corresponding to the amplification step for two different concentrations per antibiotic are presented in Fig. 6.16. The response values summarized in Table 6.3 clearly show a significant difference between positive and negative domains for the four antibiotics. The result is furthermore promising for ampicillin, which is characterized by a particularly low MRL value (4 ppb) with regard to other members of the β-lactam family.

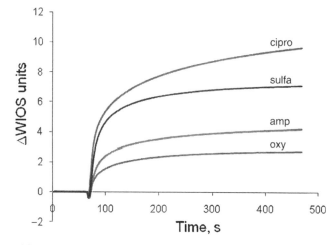

Fig. 6.16 WIOS differential response curves corresponding to amplification step of immunoassay obtained for four antibiotics: sulfapyridine (*sulfa*), ciprofloxacin (*cipro*), ampicillin (*amp*), oxytetracycline (*oxy*) at MRL concentrations: 100 ppb for sulfapyridine, ciprofloxacin, and oxytetracycline and 4 ppb for ampicillin

Blind Sample Validation

The multiplexed test developed was validated through the analysis of a series of blind milk samples provided by Nestlé. Prior to the experiment six microfluidic cartridges were prepared by filling the solution reservoirs and sample vials containing the immunoassay reagents were stored at 4°C. The tests were carried out following the experimental protocol described in the section "Test Protocol." For each antibiotic family, the differential WIOS signal was converted into a binary response: positive or negative. Afterwards, the test results for the panel of blind samples were analyzed based on the known spiked concentration values, as summarized in Fig. 6.17. It is important to notice that the unusual lack of response for the β-lactam test was surely due to the fading of activity of the corresponding competitive reagents. However, the multiplexed automated detection has been efficient for sulfonamides, fluoroquinolones, and tetracyclines. For those three antibiotic families the test accuracy reached 95%, with only one false negative observed in the case of fluoroquinolones.

Conclusions

A fully automated LOC system for simultaneous detection of multiple antibiotic families in raw milk has been successfully developed. This achievement emphasizes the relevance of a multidisciplinary approach that embraces biosensing technology

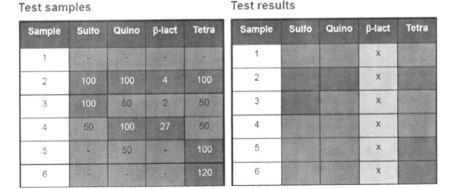

Fig. 6.17 Summary of binary responses, positive (*red*) or negative (*green*), obtained during validation test (*right table*) compared with antibiotic content of milk samples in ppb (*left table*). For each blind milk sample a single-use cartridge was used

for multiplexed detection, microfluidic cartridge design and fabrication based on new polymer materials, and system automation via appropriate fluidic setup. Individual immunoassays for specific families of antibiotics were implemented on a WIOS detection system allowing simultaneous detection. The design and fabrication of a polymer-based microfluidic cartridge using a combination of micromachining and UV micromolding with an acrylated HBP make the test cheap, fast, and easy to perform. The actuation of the disposable and passive self-contained microfluidic cartridge is performed via a simple fluidic setup involving concerted single pump/valve operation. During storage, solution mixing between reservoirs is prevented by air plugs that are deviated to loading waste during the measurement, resulting in continuous liquid delivery on the sensor chip. Sample introduction relies on the use of disposable vials with septum caps that are easily plugged into the fluidic loop setup after filling. The multiplexed immunoassay test carried out for sulfonamides, fluoroquinolones, β-lactams, and tetracyclines draws out a clear distinction between positive and negative test domains, even for ampicillin, which exhibits low MRL values. On this basis, the LOC format test has been successfully validated by analyzing qualitatively ten blind samples provided by Nestlé. The present LOC-based detection system opens up a wide spectrum of applications in the fields of food analysis, environmental monitoring, and medical diagnosis.

References

1. Jansen WTM et al (2006) Bacterial resistance: a sensitive issue: complexity of the challenge and containment strategy in Europe. Drug Resist Updat 9:123–133
2. Liu W, Zhang Z, Liua Z (2007) Determination of β-lactam antibiotics in milk using micro-flow chemiluminescence system with on-line solid phase extraction. Anal Chim Acta 592:187–192

3. Turnipseed SB et al (2008) Multi-class, multi-residue liquid chromatography/tandem mass spectrometry screening and confirmation methods for drug residues in milk. Rapid Commun Mass Spectrom 10:1467–1480

4. http://www.dsm.com/en_US/html/dfsd/tests.htm

5. Adrian J et al (2008) A multianalyte ELISA for immunochemical screening of sulfonamide, fluoroquinolone and ß-lactam antibiotics in milk samples using class-selective bioreceptors. Anal Bioanal Chem 391:1703–1712

6. Knecht BG et al (2004) Automated microarray system for the simultaneous detection of antibiotics in milk. Anal Chem 76:646–654

7. http://www.twinsensor.com

8. Wiki M, Kunz RE (2000) Wavelength-interrogated optical sensor for biochemical applications. Opt Lett 25(7):463–465

9. Cottier K et al (2003) Label-free highly sensitive detection of (small) molecules by wavelength interrogation of integrated optical chips. Sens Actuators B Chem 91:241–251

10. Adrian J et al (2009) Waveguide interrogated optical immunoSensor (WIOS) for detection of sulfonamide antibiotics in milk. Biosens Bioelectron, doi:10.1016/j/bios.2009.04.036

11. Tiefenthaler K, Lukosz W (1989) Sensitivity of grating couplers as integrated-optical chemical sensors. J Opt Soc Am B 6:209–220

12. Dübendorfer J, Kunz RE (1998) Compact integrated optical immunosensor using replicated chirped grating coupler sensor chips. Appl Opt 37:1890–1894

13. Adrian J et al (2008) Label-free waveguide simultaneous detection of sulfonamide, fluoroquinolone and β-lactam antibiotics in milk using a wavelength interrogated optical system (WIOS). Euroresidue VI: Conference on Residues of Veterinary Drugs in Food, The Netherlands, 19–21 May 2008, p 885

14. Metz S et al (2004) Polyimide microfluidic devices with integrated nanoporous filtration areas manufactured by micromachining and ion track technology. J Micromech Microeng 14:324–331

15. Junji F, Yusuke S, Kohji N (2006) Novel hepatocyte culture system developed using microfabrication and collagen/polyethylene glycol microcontact printing. Biomaterials 27:1061–1070

16. Howell Jr PB et al (2008) Two simple and rugged designs for creating microfluidic sheath flow. Lab Chip 8:1097–1103

17. Lin Y-C et al (2004) Experimental investigation of a microfluidic driving system for bi-directional manipulation. Sens Actuators A Phys 112:142–147

18. Tsao CW et al (2007) Low temperature bonding of PMMA and COC microfluidic substrates using UV/ozone surface treatment. Lab Chip 7:499–505

19. Tan HY et al (2008) A lab-on-a-chip for detection of nerve agent sarin in blood. Lab Chip 8:885–891

20. Brown L et al (2006) Fabrication and characterization of poly(methylmethacrylate) microfluidic devices bonded using surface modifications and solvents. Lab Chip 6:66–73

21. Yussuf AA et al (2005) Microwave welding of polymeric-microfluidic device. J Micromech Microeng 15:1692–1699

22. Ramser K et al (2005) A microfluidic system enabling Raman measurements of the oxygenation cycle in single optically trapped red blood cells. Lab Chip 5:431–436

23. Tourovskaia A, Figueroa-Masot X, Folch A (2005) Differentiation-on-a-chip: a microfluidic platform for long-term cell culture studies. Lab Chip 5:14–19

24. Doh Il, Cho Young-Ho (2005) A continuous cell separation chip using hydrodynamic dielectrophoresis (DEP) process. Sens Actuators A Phys 121:59–65

25. Li JM et al (2008) PMMA microfluidic devices with three-dimensional features for blood cell filtration. J Micromech Microeng 18:095021

26. Khademhosseini A et al (2004) Molded polyethylene glycol microstructures for capturing cells within microfluidic channels. Lab Chip 4:425–430

27. Witek MA et al (2008) 96-Well polycarbonate-based microfluidic titer plate for high-throughput purification of DNA and RNA. Anal Chem 80:3483–3491

28. Ye M-Y, Yin X-F, Fang Z-L (2005) DNA separation with low-viscosity sieving matrix on microfabricated polycarbonate microfluidic chips. Anal Bioanal Chem 381:820–827

29. Hong C-C, Choi J-W, Ahn CH (2004) A novel in-plane passive microfluidic mixer with modified Tesla structures. Lab Chip 4:109–113
30. Ahn CH et al (2004) Disposable smart lab on a chip for point-of-care clinical diagnostics. Proc IEEE 92:154–173
31. Mezzenga R, Boogh L, Månson J-AE (2001) A review of dendritic hyperbranched polymer as modifiers in epoxy composites. Compos Sci Technol 61:787–795
32. Eom Y et al (2002) Internal stress control in epoxy resins and their composites by material and process tailoring. Polym Compos 23:1044–1056
33. Klee JE et al (2001) Hyperbranched polyesters and their application in dental composites: monomers for low shrinking composites. Polym Adv Technol 12:346–354
34. Kou HG, Asif A, Shi WF (2003) Hyperbranched acrylated aromatic polyester used as a modifier in UV-curable epoxy acrylate resins. Chin J Chem 21:91–95
35. Inoue K (2000) Functional dendrimers, hyperbranched and star polymers. Progr Polym Sci 25:453–571
36. Rodlert M et al (2004) Hyperbranched polymer/montmorillonite clay nanocomposites. Polymer 45:949–960
37. Mourey TH et al (1992) Unique behavior of dendritic macromolecules: intrinsic viscosity of polyether dendrimers. Macromolecules 25:2401–2406
38. Luciani A et al (2004) Rheological and physical properties of aliphatic hyperbranched polyesters. J Polym Sci B Polym Phys 42:1218–1225
39. Boogh L, Pettersson B, Månson J-AE (1999) Dendritic hyperbranched polymers as tougheners for epoxy resins. Polymer 40:2249–2261
40. Lorenz H et al (1998) High-aspect-ratio, ultrathick, negative-tone near-UV photoresist and its applications for MEMS. Sens Actuators A Phys 64:33–39
41. Löchel B et al (1995) Magnetically driven microstructures fabricated with multilayer electroplating. Sens Actuators A Phys 46:98–103
42. Conédéra V, Le Goff B, Fabre N (1999) Potentialities of a new positive photoresist for the realization of thick moulds. J Micromech Microeng 9:173–175
43. Williams JD, Wang W (2004) Study on the postbaking process and the effects on UV lithography of high aspect ratio SU-8 microstructures. J Microlithogr Microfabrication Microsyst 3:563–568
44. Lee LP et al (1998) High aspect ratio polymer microstructures and cantilevers for bioMEMS using low energy ion beam and photolithography. Sens Actuators A Phys 71:144–149
45. Varadan VK, Varadan VV (2000) Micro pump and venous valve by micro stereo lithography. Proc SPIE Smart Struct Mater 3990:246–254
46. Varadan VK, Varadan VV (2001) Micro stereo lithography for fabrication of 3D polymeric and ceramic MEMS. Proc SPIE 4407:147–157
47. Schift H (2008) Nanoimprint lithography: an old story in modern times? A review. J Vac Sci Technol B Microelectron Nanometer Struct 26:458–480
48. DeSimone J, Rolland J, Denison G (2005) Functional materials and novel methods for the fabrication of microfluidic devices. WO/2005/084191
49. Rolland J et al (2004) Functional perfluoropolyethers as novel materials for microfluidics and soft lithography. Abst Pap Am Chem Soc 228:U386–U386
50. Schmidt LE et al (2007) Time-intensity transformation and internal stress in UV-curable hyperbranched acrylates. Rheol Acta 46:693–701
51. Jin Y-H et al (2007) A fast low-temperature micromolding process for hydrophilic microfluidic devices using UV-curable acrylated hyperbranched polymers. J Micromech Microeng 17:1147–1153
52. Magnusson H, Malmström E, Hult A (1999) Synthesis of hyperbranched aliphatic polyethers via cationic ring-opening polymerization of 3-ethyl-3-(hydroxymethyl)oxetane. Macromol Rapid Commun 20:453–457
53. Jin Y-H et al (2007) A fast low-temperature micromolding process for hydrophilic microfluidic devices using UV-curable acrylated hyperbranched polymers. J Micromech Microeng 17:1147–1153

54. Schmidt LE et al (2008) Acrylated hyperbranched polymer photoresist for ultra-thick and low-stress high aspect ratio micropatterns. J Micromech Microeng 18:045022
55. Nie F-Q, Macka M, Paull B (2007) Micro-flow injection analysis system: on-chip sample preconcentration, injection and delivery using coupled monolithic electroosmotic pumps. Lab Chip 7:1597–1599
56. Nie F-Q et al (2007) Robust monolithic silica-based on-chip electro-osmotic micro-pump. Analyst 132:417–424
57. Schmidt L et al (2007) Hyperbranched polymer for micro devices. WO/2007/054903
58. Natali M et al (2008) Rapid prototyping of multilayer thiolene microfluidic chips by photopolymerization and transfer lamination. Lab Chip 8:492–494
59. Kim SH et al (2007) Simple route to hydrophilic microfluidic chip fabrication using an ultraviolet (UV)-cured polymer. Adv Funct Mater 17:3493–3498
60. Jeong HE, Suh KY (2008) On the role of oxygen in fabricating microfluidic channels with ultraviolet curable materials. Lab Chip 8:1787–1792
61. Zhou WX, Chan-Park MB (2005) Large area UV casting using diverse polyacrylates of microchannels separated by high aspect ratio microwalls. Lab Chip 5:512–518

Chapter 7
Nanowire Development and Characterization for Applications in Biosensing

Robert MacKenzie, Vaida Auzelyte, Sven Olliges, Ralph Spolenak, Harun H. Solak, and Janos Vörös

Introduction

A nanowire is an extremely thin wire with a diameter on the order of a few nanometers and with lengths orders of magnitude larger than its diameter. The physical properties of nanowires at this scale are expected to deviate significantly from the bulk metal, due to confinement and surface effects. For example, the electrical conductivity of the wires changes considerably, due to the drastic increase in the surface-to-volume ratio, which can be exploited for sensing. Mechanical properties, such as the yield strength, are important parameters that need to be characterized for applications like flexible circuits. In order to study the nanowire properties one needs to arrange them on a surface in a controlled way.

Different metallization processes are used to create metal nanowires by the research community. The methods include lithography, self-assembly, and template-based methods [1]. However, fabrication of well-ordered nanostructures with sub-100-nm-range resolution, with the required quality and quantity, is a challenge for conventional fabrication techniques. Sub-10-nm-range nanostructures are especially difficult to fabricate, and only several of these methods have been reported to yield nanowires below the size of 10 nm. Electron beam lithography (EBL) was used to make NiCr nanowires as small as 3 nm, although this technique has typically limited packing density, small exposure area, and low throughput [2]. Innovative fabrication concepts, such as superlattice nanowire pattern transfer (SNAP), uses a GaAs/AlGaAs template to directly transfer metal lines onto a substrate, with which 8-nm Pt nanowires have been produced [3]. Other 10-nm-range Au and Pt nanowires have also been made using aligned block–copolymer nanostructures when loaded with anionic metal solutions and treated with oxygen plasma [4].

Janos Vörös (✉)
ETHZ – Swiss Federal Institute of Technology, Zürich, Switzerland

G. De Micheli et al., *Nanosystems Design and Technology*,
DOI 10.1007/978-1-4419-0255-9_7, © Springer Science+Business Media LLC 2009

Extreme Ultraviolet Interference Lithography

Extreme ultraviolet interference lithography (EUV-IL) is a newly emerging nano-lithography method that combines the advantages of a parallel fabrication process with high resolution. These features make it an attractive tool for researchers who are increasingly in need of nanopatterning capability that is beyond what is available from other methods such as photolithography, EBL, and scanning probe lithography, in terms of resolution or throughput.

In the majority of reported EUV-IL work, the wavelength of choice has been 13.4 nm. The reason behind this choice is twofold. First, the next-generation EUV lithography (EUVL) technology [5] that is being developed for fabrication of future semiconductor devices uses this wavelength. The EUV-IL method has played an important role in the development of high-resolution resists for this technology by making high-resolution exposures available long before projection tools become available [6, 7]. Second, it is relatively easy to fabricate diffraction gratings with sufficiently high diffraction and transmission efficiency at this wavelength on silicon nitride membranes. Finally, the short inelastic free path length of photoelectrons created by EUV photons at this energy (92.5 eV) means that the blurring effect of photoelectrons will be insignificant down to sub-10-nm-level resolution. We should also mention that this wavelength allows an ultimate half-pitch resolution of below 4 nm, which can be as small as a quarter of the wavelength.

The EUV-IL system is part of the X-ray interference lithography (XIL) beamline of the Swiss Light Source (SLS) [8]. The basic interference scheme that is used at the Paul Scherrer Institute (PSI) system is illustrated in Fig. 7.1. The spatially

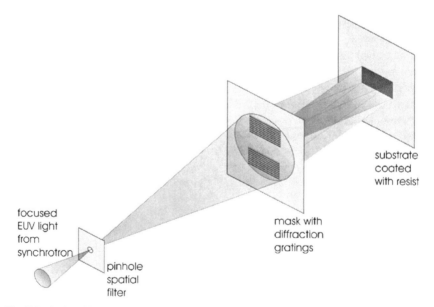

Fig. 7.1 A simplified schematic description of EUV-IL lithography. The synchrotron light is diffracted at gratings created by either e-beam lithography or EUV-IL itself

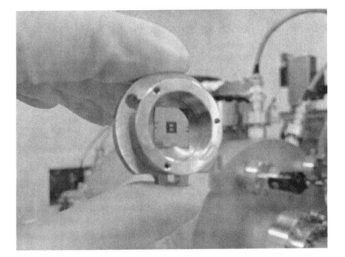

Fig. 7.2 An EUV-IL mask used to expose lines in an area of 0.5 mm²

coherent incident beam is diffracted by linear gratings that are patterned on a single partially transparent membrane. The resulting mutually coherent beams interfere at a certain distance along the beam direction. The interference fringes are used to expose a pattern in a resist film. The scheme in Fig. 7.1 shows how a periodic pattern that is two times smaller than that of the interferometer grating is created. Pairs of diffraction gratings with different periods can be placed on a single mask and printed simultaneously in one exposure step. The size and the shape of the whole printed pattern are the same as the grating size and shape. An example of the grating is shown in Fig. 7.2. In this way patterns from the micrometer to the nanometer scale can be created in a single exposure.

Metal Nanowire Fabrication

The goal of our fabrication effort is to produce nanowire arrays with periods of less than 100 nm over areas on the order of several square millimeters. It is important to be able to produce a substantial number of samples to allow experimentation. The EUV-IL method is a suitable method to achieve this goal with a demonstrated resolution of 12.5 nm in terms of the half-pitch of the periodic structures [9]. The fabrication with this method starts with the coating of the substrate with a radiation-sensitive film (photoresist), which is exposed to an EUV interference field at the XIL beamline of the SLS. The photoresist is then developed to produce an array of parallel lines. The example of the resist pattern is shown in Fig. 7.4e. To produce metal nanowires, this pattern has to be transferred to a metal film.

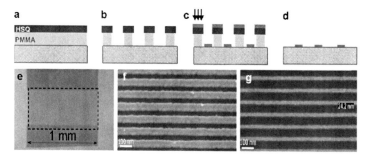

Fig. 7.3 An example of a process using a double resist layer. **a** The bilayer of HSQ/PMMA is exposed with an EUV-based pattern. **b** After the development of HSQ, the underlying PMMA layer is etched in oxygen. **c** Au film is deposited at normal incidence. **d** The final lift-off step in acetone leaves a wire array on the substrate. **e** Optical micrograph of an EUV-IL exposed field with a size of $1 \times 0.5\,mm^2$, where the nanowire area is marked with the dashed frame in the center. SEM images of 100-nm-period Au nanowires with widths of 70 nm (**f**) and 25 nm (**g**) are the result after the lift-off step

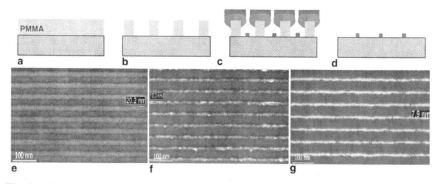

Fig. 7.4 The shadow deposition/evaporation process, where the *lines* are patterned in a single layer of PMMA and modified using shadow evaporation before lift-off. **a, b** PMMA is spin coated, exposed, and developed. **c** The shadow deposition of Cr is followed by normal deposition of Cr and/or Au. **d** The entire stack is lifted off. **e** 50-nm-period PMMA lines after development. **f, g** SEM images of resulting 50-nm-period 10-nm Cr and 8-nm Au nanowires, respectively

For the fabrication of the metal nanowires we developed a number of modified lift-off processes, such as the use of two photoresists in a bilayer stack and shadow evaporation of metal layers to modify the width and profile of photoresist lines [10]. The processes are shown in Figs. 7.3 and 7.4. The bilayer stack approach provides an undercut that is needed for the following lift-off step to work. The oblique-incidence evaporation process provides robust control over the line width through the shadowing effect. The technique is limited to patterns of certain geometry, such as periodic arrays of lines or dots. Obtaining line widths in the 5- to 10-nm range through a lift-off process directly with a resist pattern requires gaps (spaces) of this size, which is very difficult to achieve because available resists usually lack sufficient resolution at this scale. Such narrow gaps in a resist are especially difficult to achieve in interference lithography because of the sinusoidal profile of the aerial image.

Using these processes we were able to obtain metal lines with widths in the 10 to 70-nm range. The wires were made on silicon and polyimide substrates. Smooth and regular 100-nm-period, 70- and 25-nm-wide Au lines are demonstrated in Fig. 7.3f, g. Using PMMA width of 50 nm with these processes it was possible to obtain dense, 10 and 8 nm wide chromium and gold nanowires (Fig. 7.4f, g). At this scale, the grain structure and growth properties of the metal during deposition become critical, causing roughness and, ultimately, continuity problems. Control of parameters such as substrate surface chemistry, deposition rate, pressure, and temperature and the metal (or alloy) type is important for obtaining high-quality nanowires.

Nanowire Mechanical Characterization

In this module the mechanical properties of gold nanowires were investigated by in situ tensile testing at the SLS. Mechanical properties were determined as a function of temperature and strain rate. The content of this section has been published in three peer-reviewed papers, one in *Acta Materialia* and two in *Scripta Materialia* [11–13]. The basis for an understanding of mechanical properties is a detailed knowledge of the microstructure of a material in the specific case of gold nanowires, which constitutes the starting point of our study. As the testing of nanostructures is quite challenging, a subsection is dedicated to the testing method. The results are discussed in terms of small-scale plasticity, strain gradient plasticity, diffusional flow, and fracture mechanics.

Microstructure

The most interesting aspects of the nanowire microstructure are (a) the lines exhibit a strong (111) fiber texture, as one would expect from thin metallic fcc films, (b) the grain size is about half the wire width, and (c) the lines exhibit a very low edge roughness, which is a sign of high lithographical quality. Details can be observed in Fig. 7.5.

Mechanical Testing

The line arrays were tested at the Materials Science Beamline at the SLS by an in situ technique, which was originally developed for tensile testing of metal thin films by Bohm et al. [14]. This technique measures the changes in lattice strain as a function of external load and is described in detail in [14]. Sample load is generally associated with changes in lattice strains, which are observed by X-ray diffraction. The energy of the X-ray beam has to be adjusted to the sample texture in order to

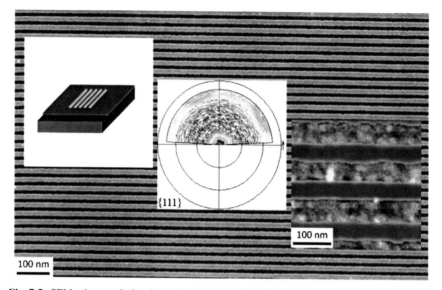

Fig. 7.5 SEM micrograph showing microstructure of nanolines (sample E): the *lines* show grains in a range of 20 to 30 nm. The *inset* shows the {111} pole figure of the lines obtained by XRD. The peak in the center and the continuous ring at about 70° indicate a (111) texture. In addition, the continuous ring is superimposed by a weak fourfold symmetry, which is caused by a weak sidewall texture

obtain a Debye–Scherrer ring for normal beam incidence. Due to sample symmetry it is convenient to observe the changes in lattice strain for the principal axis of the Debye–Scherrer ring that correspond to the load $\varepsilon_L(N)$ and the transverse $\varepsilon_T(N)$ direction, where N is an integer number and labels the external load level applied by the tensile tester to the sample. The detector signal for the principal axes of the Debye–Scherrer ring is shown in Fig. 7.6. The lattice strain is calculated in terms of principal strains $\varepsilon_1(N)$, $\varepsilon_2(N)$ and subsequently converted into principal stresses $\sigma_1(N)$, $\sigma_2(N)$ by Hooke's Law. It is important to note that $\varepsilon_3(N)$ depends linearly on $\varepsilon_1(N)$ and $\varepsilon_2(N)$ due to the boundary condition $\sigma_3 = 0$ for nonpassivated samples. External strains ε are measured by tracking two markers close to the parallel line ensemble using a LaVision photographical system. At least one micrograph is taken for each external load level N. Data are analyzed using a MathLab routine developed by Eberl et al. [15]. This allows for a "local" determination of the strain $\varepsilon(N)$ as a function of the external load level N.

Finally, the experimental data of strain $\varepsilon(N)$ and stresses $\sigma_1(N)$, $\sigma_2(N)$ are merged to "full" stress–strain curves $\sigma_1(\varepsilon)$ and $\sigma_2(\varepsilon)$. It is important to consider that this technique measures relative stresses and not absolute stresses. This technique can be transferred fully from thin films to our parallel line arrays. The only important condition is a nonpassivated, free surface of the samples in order to satisfy the assumption $\sigma_3 = 0$ (cf. [14]), which is the case in our experiment. Since the parallel nanoline arrays show the same (111) texture like the thin films of Bohm

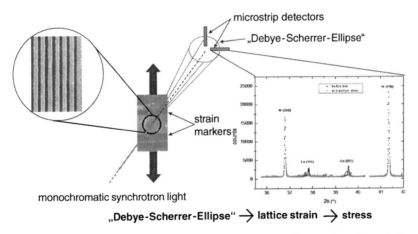

Fig. 7.6 Experimental setup at SLS-MS Beamline. The two second-generation microstrip detectors are mounted to measure the principal axis of the Debye–Scherrer ring (load direction and transverse direction) for normal beam incidence

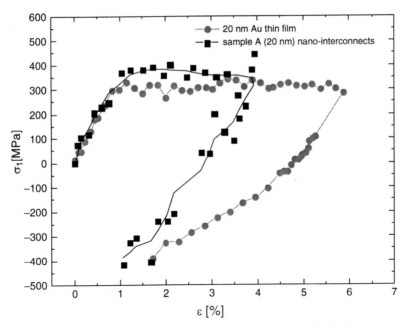

Fig. 7.7 Representative stress–strain behavior of parallel line arrays of gold for a continuous gold thin film and a line array from sample A. For better illustration, the original data have been smoothed by a three-point fast Fourier transformation (*line*)

et al. [14] (cf. inset of Fig. 7.7), the explicit formalism for textured samples can be used without any modification.

The tensile tester (Kammrath and Weiss) was mounted on the $x/y/\varphi$ powder diffractometer of the Materials Science Beamline (MS Beamline) [16]. Defining a left-handed Cartesian coordinate system, the X-ray beam is in the x-direction, the load axis is in the z-direction, and the parallel line array is in the y/z plane (where the nanointerconnects are aligned in the z-direction). The X-ray beam was adjusted to the texture of the gold lines and had an energy of 7.97 keV. Two second-generation microstrip detectors (line detectors) [17,18] are mounted in the x/z plane (load direction) and in the x/y plane (transverse direction) for monitoring changes in lattice strain. The distance and position of the detectors are chosen to observe diffraction angles of about $38 \pm 4°$ corresponding to the diffraction angle of the peak of gold for the used X-ray energy.

A thin film of tungsten powder suspended in vacuum grease is applied on the reverse side of the Kapton™ substrate. The positions of the tungsten peak and the unstressed gold peak (image number $N = 1$) are assigned to their corresponding 2θ angles for calibrating the detectors. Since no stress is transferred from the vacuum grease to the tungsten powder, the tungsten peak is set to the corresponding 2θ angle for all external load states N. Gold peak positions are calculated relative to the well-defined tungsten peak and the calibration of the detector. Possible shifts of the tungsten peak are due to small changes in the experimental setup, e.g., sample sliding or sample movement in the beam direction.

Results

Figure 7.7 shows a typical stress–strain curve of a 20-nm nanowire array. It hardly exhibits any strain hardening. The tensile properties of the nanolines were also monitored as a function of temperature, which was adjusted in the range of 173–373 K using a CryoJet (Oxford Instruments) available at the SLS-MS beamline. The corresponding results are shown in Fig. 7.8. Starting with an elastic slope, the stress in the load direction reaches an upper plateau. After unloading, the stress follows again a linear elastic regime till a lower plateau is reached. The level of the upper plateau is defined as the yield strength of the gold nanowires and strongly depends on the temperature during testing. However, this behavior is unexpected for fcc materials such as gold, which does not show temperature-dependent mechanical properties below homologous temperatures of 0.5 for the bulk regime.

Discussion

Several important topics are considered in this section. First, the discussion focuses on the yield stresses and the residual stress. Afterward, cracking and delamination of the nanolines are analyzed by a theory of Xia et al. [19], which regards the strain transfer from flexible substrates to the line arrays. Finally, the yield stresses of the nanointerconnects are compared with yield stresses of thin films obtained by tensile testing [20] as well as yield stresses of nanowires obtained by bending experiments [21]. For the latter, the introduction of strain gradient plasticity theory is required.

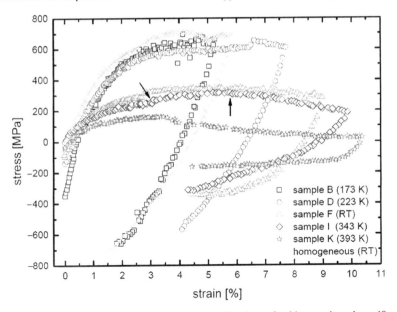

Fig. 7.8 Representative stress–strain curves (loading direction) of gold nanowires about 40 nm in width, 20 nm in in height, and 1 mm in length for different testing temperatures T_t. The *vertical arrow* marks the point of doubling the strain rate, and the *sloping arrow* marks the onset of cracking for sample *I*

The measurement of residual stresses σ_r at the synchrotron was not possible due to time constraints and a lack of all required degrees of freedom of the goniometer. As the residual stress σ_y is an important quantity, three strategies were applied for its determination: the postulation of a symmetric yield behavior in tension and compression, measurements of residual stresses on blanket films deposited in the proximity of the nanowires, and an estimation of residual stress from the deposition conditions.

In general, these samples show plateaus at σ_{yt} (tensile stress) and at σ_{yc} (compressive stress), where $|\sigma_{yc}| \neq |\sigma_{yt}|$. The FWHM of the X-ray peaks as a function of strain indicates that no dislocation networks are formed, which excludes a Bauschinger effect for explaining the nonequality of $|\sigma_{yc}|$ and $|\sigma_{yt}|$. Consequently, the difference in absolute value of σ_{yt} and σ_{yc} has to be caused by another asymmetric effect, which is assumed to be the residual stress σ_r. To obtain the expected plateaus at $\pm\sigma_y$, the stress–strain curves have to be shifted by

$$-\frac{\sigma_{yt} + \sigma_{yc}}{2} = \sigma_r. \tag{7.1}$$

For $\sigma_{yt} = 450$ MPa and $\sigma_{yc} \approx -320$ MPa (sample A), a residual stress of $\sigma_r = -65$ MPa is calculated by (7.1).

This estimation assumes that the residual stress is the only asymmetric effect in the nanolines. To strengthen this assumption we present some additional arguments that confirm the value of residual stress obtained by (7.1).

First, the residual stress σ_r is measured by $\sin^2(\Psi)$ experiments [22] using common XRD systems (Eulerian cradle and monocapillary with 300-μm spot size). The residual stress is measured on homogeneous pads present on the same Kapton wafer and that have a deposition history and microstructure identical to those of the nanowires. The XRD configuration for measuring the residual stress using the (111) peaks, which are required because of the small scattering volume of the sample, is very sensitive to small misalignments in sample height. A calibration (BaO$_2$ powder suspended in vacuum grease) is applied on the top of the samples to minimize this undesirable effect. Two different XRD lab systems are used to avoid systematic alignment errors. However, all measurements yield results in the range of -40 to -110 MPa, which is in good agreement with the residual stress σ_r of -65 MPa obtained by (7.1).

The residual stress can be verified by another estimation using the differences in thermal expansion of the polyimide substrate and gold line arrays. The effective stiffness (Young's modulus times thickness) of polyimide exceeds the stiffness of a referential homogeneous 30-nm gold thin film (corresponding to the maximum height of the line arrays) by two orders of magnitude. Consequently, the strain of the line arrays is given by the substrate. In this frame, the residual stress is given by

$$\sigma_r \approx \left(\alpha_{\text{Kapton}} - \alpha_{\text{Au}}\right) E_{\text{Au}} \, \Delta T, \qquad (7.2)$$

where α_{Kapton} and α_{Au} are the thermal expansion coefficients of polyimide and gold, respectively, $\Delta T = T_0 - T_d$ is the temperature difference between room temperature T_0 and deposition temperature T_d, and E_{Au} is Young's modulus of gold. For $\alpha_{\text{Kapton}} = 30 \times 10^{-6}\,\text{K}^{-1}$,[1] $\alpha_{\text{Au}} = 14.2 \times 10^{-6}\,\text{K}^{-1}$, $E_{\text{Au}} = 80$ GPa [corresponding to the (110) in-plane isotropy for (111) textured samples], and the above estimated residual stress of $\sigma_r = -65$ MPa, an increased temperature for gold deposition of about 50 K with respect to room temperature is obtained, which is reasonable.

The exclusion of a Bauschinger effect may be contrary to the observations of Xiang et al. [23,24]. However, the formalism described by Xiang et al. [24] is based on the interaction of numerous dislocations per grain that are indeed present in the described thin films in the thickness range of several hundred nanometers (submicron range) to several microns. In comparison, the thickness as well as the average grain size of the parallel line arrays in the current study are at least one order of magnitude smaller. Consequently, less than one dislocation per grain on average is expected for the nanointerconnects, if a similar dislocation density is assumed. Hence, no dislocation networks are expected to be formed at the sample/substrate

---

[1] Thermal expansion of Kapton^{TM} is numbered in the range of 30–60 $\times$ 10^{-6} K^{-1}. From experience, the Kapton^{TM} foils show rather the lower value of 30 $\times$ 10^{-6} K^{-1} which is used for further calculations.

interface, which is commonly quoted as the origin of a Bauschinger effect in thicker films [24].

The yield stresses σ_y are obtained by the intersection of a line parallel to the elastic slope at $\varepsilon = 0.5\%$ and $\sigma_1(\varepsilon)$. For example, samples A and B show yield stresses of 450 and 330 MPa, respectively. As a consequence of residual stress, the stress–strain behavior must be corrected by the above estimated residual stress σ_r of -65 MPa resulting in "real" yield stresses of 385 and 265 MPa, respectively.

Strain transfer from substrates to thin films is discussed in detail by Xia et al. [19]. They report characteristic length scales l for transferring stresses from substrates to thin films depending on Dundur's parameters α and β [25] defined by

$$\alpha = \frac{\overline{E_1} - \overline{E_2}}{\overline{E_1} + \overline{E_2}} \text{ and } \beta = \frac{\mu_1(1 - 2v_2) - \mu_2(1 - 2v_1)}{2\mu_1(1 - v_2) + 2\mu_2(1 - v_1)}, \quad (7.3)$$

where $\overline{E_i} = E_i/(1 - v_i)$ are the plane strain tensile moduli, μ_i are the shear moduli, E_i are Young's moduli, and v_i are the Poisson ratios of the sample ($i = 1$) and the substrate ($i = 2$). These parameters concern the mismatch in elastic properties between substrate and sample materials. The length scale l for full stress transfer is given by [19]

$$l = \frac{\pi}{2} g(\alpha, \beta) h, \quad (7.4)$$

where $g(\alpha, \beta)$ is a function related to Beuth [26] and h is the film thickness. Since $g(\alpha, \beta)$ depends strongly on α and only weakly on β, we neglect the weak dependence on β for further consideration. For $E_{Au} = 80$ GPa, $v_{Au} = 0.57$ [in-plane for (111) texture], $E_{Kapton} = 3$ GPa, and $v_{Kapton} = 0.34$, a Dundur parameter of $\alpha = 0.94$ is obtained, corresponding to $g(\alpha, \beta) \approx 10$ and a characteristic length scale l of approx. $15h$ [19]. For sample heights h of 20 to 30 nm, a characteristic length l for full strain transfer (from Kapton to the parallel line array of gold) of less than 500 nm is expected. Since the length of the line arrays (1 mm) exceeds this value by three orders of magnitude, this fringe effect should be negligible for ductile materials like gold.

Nevertheless, if samples show stress relaxation (tensile as well as compressive stress relaxation), e.g., due to poor adhesion of the lines to the substrate, the lines are expected to crack at points of maximum strain. Thus, crack distances in the range of $2l$ are expected. This is consistent with the observations of Gruber et al. [27] for Ta/Cu/Ta multilayer thin films. In reference to Figs. 7.7 and 7.9, sample A does not show obvious evidence of delamination. In contrast, sample B shows a clear decrease of about 100 MPa after passing the maximum stress at strains of 1.5 to 2%, indicating tensile stress relaxation due to line cracking or delamination of the lines [14]. The delamination of the lines is most probably caused by an insufficient adhesion of the lines. This can be confirmed by posttesting SEM micrographs, which show delamination/buckling lines perpendicular to the nanointerconnects. The distances of these delamination/buckling lines are in the range of 1 to 2 μm, which is in good agreement with the above estimated distance of $2l \approx 1$ μm. Since it is assumed that the early delamination of sample B at strains of about 1.5 to

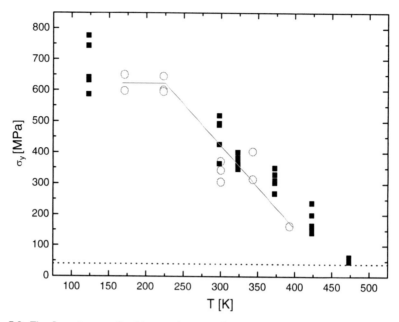

Fig. 7.9 The flow stresses of gold nanowires as a function of temperature (*open circles*) and the flow stresses of Au thin films with thicknesses between 80 and 500 nm (*black squares*) for comparison. An athermal regime can be observed below 220 K

2% influences the yield stress, this sample shows rather a "delamination threshold stress" than a typical yield stress (which is remarkably lower in comparison to the yield stress of sample A).

The "real" stress–strain behaviors of the nanointerconnects are compared to the behaviors of gold thin films that were obtained by the same tensile technique [20]. The nanointerconnects show, in good agreement, the same stress–strain behavior as homogeneous gold thin films of equal thickness. From this we conclude that nanolines and thin films have comparable grain sizes, which are assumed to be the governing strengthening mechanism [20].

On the other hand, the obtained maximum yield stresses of about 350 MPa are lower by about one order of magnitude compared to the yield stresses obtained by bending experiments for polycrystalline gold nanowires [21]. Strain gradient plasticity may give an explanation for this discrepancy. In this theory, deformations that include bending moments like torsion or bending require additional geometrical dislocations [28, 29]. The additional dislocations give a contribution to the yield stress via Taylor's formula [28]:

$$\Delta\sigma = \mu b \sqrt{\rho_S + \rho_G}, \tag{7.5}$$

where ρ_S is the statistical dislocation density, ρ_G is the geometrically necessary dislocation density, μ is the shear modulus, and b is the absolute value of Burger's vector. For bending experiments, Gao et al. [29] have derived an explicit formalism for the strain-gradient-induced increase in yield stress σ given by

$$\left(\frac{\Delta\sigma}{\sigma_0}\right)^2 = 1 + \frac{1}{R\,\rho_s\,b}, \qquad (7.6)$$

where R is the radius of curvature of the bending experiment and σ_0 the yield stress without strain-gradient-induced geometrical dislocations (here the geometrical dislocation density is related to the radius of curvature of the experiment). Figure 7.6 shows an ensemble of $\sigma\,(R, \rho_s = cte)$ calculated by (7.6) for $\sigma_0 = 385\,\text{MPa}$ (corresponding to our experimental results) and $b = 2.88 \times 10^{-10}\,\text{m}$ (absolute value of Burger's vector of gold). The yield stresses of bending-moment-containing experiments strongly increase with decreasing radii of bending curvature. Also the statistical dislocation density has a remarkable influence on the yield stresses: the higher the statistical dislocation density, the weaker the level of the yield stress. The upper hatched band marks the range of the results reported by Wu et al. [21].

The statistical dislocation density ρ_S in the absence of strain gradients is estimated by

$$\sigma_0 = \sigma_P + \sigma_{HP} + \sigma_T = \underbrace{\frac{2\mu}{(1-v)}\,e^{-2\pi\frac{\omega}{b}}}_{\sigma_P} + \underbrace{\frac{k_{HP}}{\sqrt{d}}}_{\sigma_{HP}} + \underbrace{\mu b\,\sqrt{\rho_s}}_{\sigma_T}, \qquad (7.7)$$

where σ_P is the Peierls tension, σ_{HP} the Hall–Petch stress, and σ_T the Taylor tension. The Peierls tension depends on Poisson's ratio v, shear stress μ, and Burger's vector b and on the deformation zone ω of a dislocation, and it is on the order of less than 1 MPa. Hence, we neglect the Peierls tension for the following considerations. The Hall–Petch stress depends on the average grain size d of the lines and the Hall–Petch coefficient k_{HP}, which can be estimated for fcc materials by [30]

$$k_{HP} \approx \frac{\mu\sqrt{b}}{15}. \qquad (7.8)$$

Assuming an absolute value of Burger's vector b of $2.88 \times 10^{-10}\,\text{m}$, a shear modulus of $\mu = 27\,\text{GPa}$ and an average grain size d of 20–30 nm (confer SEM microstructure of the gold nanolines in Fig. 7.5), a Hall-Petch stress in the regime of 170–220 MPa is obtained. For $\sigma_0 = 385\,\text{MPa}$ [28] Taylor stresses are calculated to be in the range of 165 to 215 MPa, which correspond to statistical dislocation densities ρ_S in the range of 5×10^{14} to $10^{15}\,1/\text{m}^2$. For the radius of curvature R, reasonable values in the regime of 250 nm to 1 µm [21] are assumed. The expected yield stresses for bending-moment-containing experiments of our gold nanointerconnects are calculated by (7.6) to be on the order of 2 to 1.1 GPa (cf. the lower hutched band in Fig. 7.10). This is lower than the data reported by Wu et al. [21] of 5.6 and 3.5 GPa for nanowires with diameters of 40 and 200 nm, respectively (cf. the upper hutched band in Fig. 7.10). An explanation for this discrepancy could be the

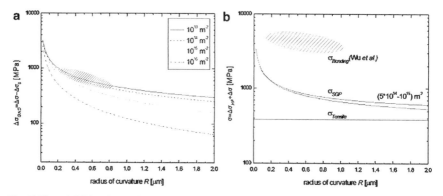

Fig. 7.10 **a** Additional contribution of geometrically necessary dislocations to yield strength as a function of the radius of curvature R, $b = 2.88 \times 10^{-10}$ m, and $\mu = 27$ GPa [cf. (7.6)]. **b** Comparison of tensile yield stress of our gold nanolines (σ_{Tensile}), the theoretical yield strength of our gold nanolines in the frame of strain gradient plasticity (σ_{SGP}), and the bending yield stress of single, freestanding gold nanowires (σ_{Bending})

considerably enhanced twin boundary density of electroplated samples compared to specimens prepared by physical vapor deposition techniques. A similar observation has been made by Shen et al. [31], who reported an additional hardening effect for electroplated copper thin films due to an increased twin boundary density. Nevertheless, our results seem to be reasonable since the theoretical yield stress of 2.7 GPa (approximated by shear modulus μ divided by 10) is expected to be observed for single crystalline gold nanolines rather than for polycrystalline ones.[2] However, even though the exact determination of the radius of curvature as well as the statistical dislocation density is difficult, Fig. 7.10 clearly shows a trend that yield stresses obtained by bending experiments are expected to increase with decreasing radii of bending curvature. This is merely a result of the bending geometry and is not caused by a scaling in intrinsic material properties obtained in uniaxial tension.

The strong temperature dependence of the yield strength is unusual for bulk fcc materials such as gold [32]. For analysis, a diffusional mechanism is chosen because of (1) the presence of free surfaces, (2) the extremely small grain size of 20 to 30 nm (short diffusion paths and the absence of stored dislocations), and (3) the very low strain rates applied. According to Gruber et al. [33], a general diffusional creep equation is assumed:

$$\dot{\sigma} = E \left(\dot{\varepsilon}_{\text{ext}} - \dot{\varepsilon}_{\text{p}} \right) = E \left(\dot{\varepsilon}_{\text{ext}} - C \sigma \right), \tag{7.9}$$

---

[2] Originally, the discrepancy between the theoretical yield stress and the experimentally observed yield stresses of polycrystalline samples led to the introduction of dislocations. In reverse, the theoretical yield strength is achievable only for dislocation-free or defect-weak specimens like single crystals or whiskers.

where E is Young's modulus, $\dot{\varepsilon}_{ext}$ is the constant external applied strain rate, $\dot{\varepsilon}_p$ is the relaxation strain rate due to diffusion-controlled plastic deformation, and C is a factor that is characteristic of the dominant diffusion mechanism. With the boundary condition $\sigma\,(\varepsilon = 0) = \sigma_0$, (7.9) can be solved by

$$\sigma\,(\varepsilon) = \frac{\dot{\varepsilon}_{ext}}{C} - \left(\frac{\dot{\varepsilon}_{ext}}{C} - \sigma_0\right)\exp\left(-E\frac{C}{\dot{\varepsilon}_{ext}}\varepsilon\right) = a + b\,\exp\left(-c\varepsilon\right). \qquad (7.10)$$

The stress–strain curves can be fitted well by (7.10) for the case $\varepsilon < \varepsilon_c$. The characteristic factor C for the diffusion mechanism generally shows a temperature dependency:

$$C \propto \frac{1}{k_B T}\,\exp\left(-\frac{E_{act}}{k_B T}\right). \qquad (7.11)$$

Here k_B is Boltzmann's constant, T is the absolute temperature, and E_{act} is the thermal activation energy. The activation energy is obtained by an Arrhenius plot of $\dot{\varepsilon}_{ext} T_t / a$ vs. $1/T_t$, which is shown in Fig. 7.11. The initial average time (pulling time plus measuring time) per applied strain step is used to calculate the strain rate $\dot{\varepsilon}_{ext}$. An exponential relationship indicating different temperature-dependent deformation mechanisms is observed in the Arrhenius plot. For temperatures above 343 K an activation energy of approx. 300 meV is obtained while for the intermediate

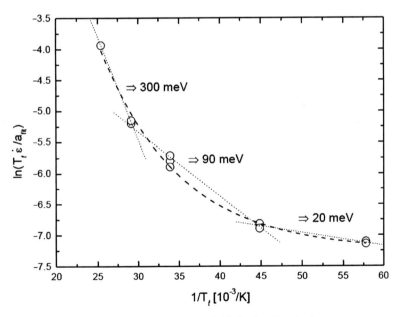

Fig. 7.11 The Arrhenius plot shows an exponential behavior. For the low temperature range, meV indicates deformation by dislocations, for the intermediate regime, meV indicates deformation by self-diffusion on free (111) surfaces, and for the high temperatures, meV indicates plastic deformation by Coble creep or grain boundary gliding

temperature range from 223 to 343 K, the data can be approximated by a linear behavior resulting in an activation energy $E_{act} \approx 90$ meV. For the low temperature regime, a nearly athermal behavior is observed. This gradual change in activation energy is interpreted by a coupling in grain-boundary and surface diffusion paths. At high temperatures, grain-boundary diffusion is dominant [33,34]. Then the influence of surface diffusion successively increases, resulting in a reduced activation energy (cf. $E_{act}^{(111)} = 120$ meV for gold (111) surface diffusion [35]). Finally, the athermal region is interpreted as a dislocation glide.

In conclusion the following points have been found to be true:

- Unpassivated gold nanowires exhibit an unexpectedly low yield strength (given their small dimensions). This is due to the fact that at such small dimensions only pure uniaxial testing reveals the true material properties. Bending geometries results in much larger apparent strengths as explained by strain gradient plasticity.
- The yield strength of gold nanowires is extremely temperature dependent and requires temperatures as low as 220 K to become athermal. This observation could be explained by diffusional flow with grain boundaries and surfaces as the dominating diffusion paths.
- The fracture toughness of gold nanowires is strongly reduced with respect to their bulk counterparts. It increases with decreasing temperature, which is due to a change in strain hardening behavior. At temperatures below 220 K a slight hardening is observed, whereas at temperatures higher than room temperature the stress–strain curves perfectly resemble elastic–plastic behavior.

Nanowire Application for Biosensing

Obviously nanowire technology is eligible and attractive for application in a diverse range of fields. The field of focus in this section, however, will remain biosensing, as well as the techniques and characterization methods required to construct biosensors using our nanowire technology. Figure 7.12 illustrates schematically the parts comprising a typical electrical biosensor.

Nanowires as Elements for Biosensing

One might think that all wires are inherently the same, but this is not true. Although all wires exist for the purpose of conducting electrons, their dimensions and material properties play an important role in how they conduct. This is especially true at the nanoscale, when their proportions shrink to a single dimension (i.e., length) and the electron transport is better described with quantum mechanical relationships, rather than classical electrical relationships. At this threshold and beyond nanowires prove

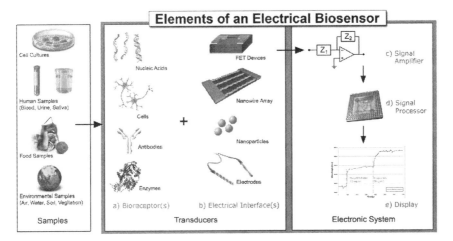

Fig. 7.12 Elements and selected components of a typical electrical biosensor. **a** Bioreceptors that specifically bind to the analyte. **b** An interface architecture where a specific biological event takes place and yields a signal, which is detected by the transducer element. **c** The transducer signal is converted into an electronic signal and amplified by a detector circuit using the appropriate reference. **d** Processing of the signal to a meaningful physical parameter. **e** The resulting quantity is presented through an interface to the human operator [36]

to be extremely interesting not only from a physics perspective, but also for their potential application in biosensing, bioelectronics, and optics.

Surface-to-Volume Characteristics

As a wire decreases in diameter to the nanometer regime, the ratio of surface atoms compared to interior atoms, i.e., the surface-to-volume ratio, drastically increases. Therefore, external influences by charged particles or biological species increasingly influence the conduction both on the wire surface and in the wire interior. Penetration into the wire interior is typically determined by the Debye screening length. Furthermore, quantum mechanical models of wavelike electron transport through a nanowire, rather than classical mechanical models of ballistic electron transport, become necessary as the diameter of nanowires approaches the order of the material-dependent Fermi wavelength [37]. This implies that conductivity in nanowires is subject to quantum confinement into separate quantized energy levels. In other words, conductance in a nanowire of a sufficiently small diameter is related to the sum of electron transport in the separate conduction channels of the nanowire defined by their quantization energy. The thinner the wire, the smaller the available unaffected volume, and hence the smaller the probability of uninterrupted conduction. Assuming that the wire is free of impurities, the conductance of each channel is described by $G = 2e^2 = h$, where e is the elemental charge of the electron and h is Planck's constant [38]. Figure 7.13 illustrates this concept, but uses the

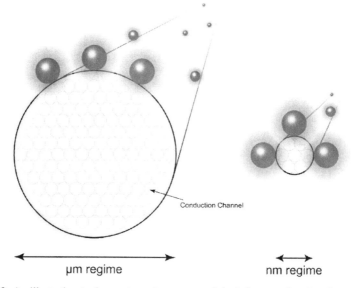

μm regime   nm regime

Fig. 7.13 An illustration to demonstrate the concept of the influence of surface interactions on nanowire conduction. In larger wires, for example with micrometer dimensions, the surface-to-volume ratio is relatively small. Even if surface interactions, such as binding with charge particles or biological species, influence conduction near the surface of the wire, there is still a large portion of the wire's interior available for uninterrupted conduction. To help visualize this, the illustration uses a mental construct of "conduction channels" within the wire. As the dimensions of a wire decrease to the nanometer regime, the surface-to-volume ratio drastically increases. Therefore, the same external influences increasingly influence the conduction both on the wire surface and in the wire interior, thereby decreasing the possibility for uninterrupted conduction

term "conduction channel" in a more classical sense as a mental construct to aid in visualization.

Nanowires, therefore, represent very attractive bioelectrochemical transducer components since their diameters are comparable to the size of the biochemical analytes under analysis and since their conductance is sensitive to surface perturbations. Nanowires have already been incorporated into FET devices for biosensing purposes, such as the detection of pH, protein and DNA binding, and viral, and cancer markers. To overcome the problems of large-scale nanowire production, we have combined EUV-IL with biological patterning to produce high-density line arrays of self-assembling DNA-tagged gold nanowires [36, 39].

The successful implementation of nanowire-based biosensing devices depends on overcoming the challenges of their precise and reproducible placement in large-scale fabrication processes, their specific biological functionalization, and their interconnection with biological and nano-/microelectronic systems. Whether the nanowires are grown, pulled, tagged and enhanced, or deposited, the accurate characterization and subsequent controlled manipulation of the electrical properties of nanowires is essential to their application-specific performance.

Nanogap Sensing

In our system, we have nanometer spaces not only between parallel wires, but also between the individual particles, of which the wires are composed (Fig. 7.16). This is a unique, interesting, and powerful property of our system, which does not inherently exist in other nanowire systems (e.g., sputtered solid nanowires). Electrically speaking, we know from our mathematical model for electron transport in particle-based systems that a gap of more than 1 nm between particles will inhibit conduction. If our goal is to obtain high conductance, then we must ensure that the particles are placed densely enough or become in some way enhanced to close the interparticle gaps. We can, however, use these gaps to our advantage. Furthermore, the controlled creation of coulomb blockades in such nanowires could enable the creation of 1-dimensional transistors [40]. Our large-scale nanowire production could allow for entire arrays of such new nanoelectronic devices.

Metallic nano-objects (particles, rods, etc.) exhibit a strong coupling of the electromagnetic fields in the interparticle gap, and the resulting characteristic optical resonance around the visible wavelength is known as localized surface plasmon resonance (LSPR). This interparticle gap is known to greatly influence the spectral shape and the optical response. This concept has already been demonstrated in dot arrays to detect single-binding events [41]. Self-assembled, particle-based nanowires have proven to be an interesting system, which could allow, for example, for the simultaneous optical and electrical observation of the system in response to stimuli.

Nanofabrication and Surface Architecture

Previously in collaboration with the PSI we had developed a platform that enables the precise positioning of DNA-tagged gold colloids onto micro- and nanopatterns (Fig. 7.14). This platform has been successfully expanded with the creation of new, more stable and robust methods for the production of even smaller nanowires and hybrid nanostructures with the use of self-assembling nanoparticles. Gold colloids of various sizes, now ranging from 5 to 100 nm or more, can be patterned onto arbitrary patterns with widths ranging from 20 nm up to several hundred microns. Our approach is unique in the sense that it can be applied from the large micron sizes down to the sub-100-nm range and that the nanopatterns cover quite a large area of a few square millimeters. Such densely packed nanowires embedded in a biopassive background are the first promising steps toward nanowire-based biosensing. We use chemical enhancement to ensure that the particle line patterns, as shown in Fig. 7.16a, b, form conducting wires (Fig. 7.17c, d). Please notice the strict separation in the definitions of the terms line (nonconducting) and wire (conducting).

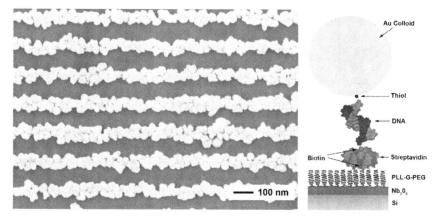

Fig. 7.14 An SEM nanowire image (after chemical enhancement) and a side-view representation of the nanoparticle attachment. Nb_2O_5-coated surfaces were patterned using the molecular-assembly patterning by lift-off (MAPL) technique [44]. This results in a biotinylated poly(L-lysine) – *graft* – poly(ethylene glycol) (PLL–*g*-PEG/PEGbiotin) foreground, surrounded by a resistant PLL–*g*-PEG background. BiotinDNA–neutravidin complexes were immobilized to these modified surfaces. Isolated gold particles tagged with complementary thiol DNA were specifically attached to the surface through the hybridization of the DNA strands

Hybridization Deposition of DNA-Tagged Au Particles

Initially, gold colloid particles 20 nm in diameter were first tagged with single-stranded DNA (ssDNA) and then deposited onto arbitrary patterns prepared with the complementary DNA. The gold particles then bound to the pattern via DNA hybridization. Figure 7.14 illustrates the biological assay resulting from this process. By tagging the gold particle with ssDNA, the resulting lines of gold particles were already well functionalized to bind with analyte molecules. A further advantage is our ability to controllably dehybridize the DNA double-helix (e.g., heat, light, salt concentrations) [42], thereby releasing targeted gold particles. This feature has also allowed us to explore DNA-assisted pattern reproduction and particle transfer techniques [43].

Unfortunately, the relatively weak attachment of the gold particles to the substrate surface made postprocessing of the pattern extremely difficult. For example, merely seconds of sonication would release the gold particles from the substrate, thereby destroying the line structure. This eliminated conventional lithographic techniques for the fabrication of contact pads and leads to the nanowire arrays.

Electrostatic Deposition of DNA-Tagged Au Particles

To enable the postprocessing of nanoline and nanowire arrays, a new method was developed that can withstand sustained high levels of sonication, as well as many steps of photolithographic exposure and development. The deposition process is

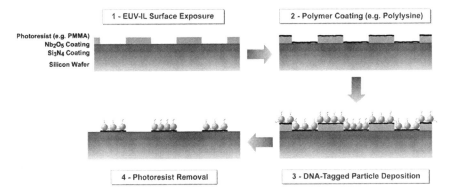

Fig. 7.15 Illustrated process for the deposition of DNA-tagged gold particles to form structured nanopatterned surfaces. This process is an extension of the MAPL surface patterning technique. Gold particles are tagged with ssDNA to enhance the overall net negative charge of the particles. These then attach strongly to positively charged polymers. The final pattern is exposed through a removal of the photoresist background

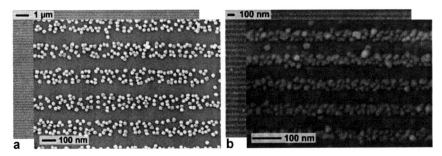

Fig. 7.16 Perspective SEM images of nonenhanced nanolines created by the self-assembly of DNA-tagged gold nanoparticles. **a** Nonenhanced 20-nm particles tagged with DNA and bound weakly via hybridization, approx. 110-nm line widths with a 200-nm period (*line/background*). **b** Nonenhanced 5-nm particles tagged with DNA bound strongly via electrostatic interaction, approx. 30-nm line widths with a 100-nm period (*line/background*), which borders on the limits of clear SEM imaging for this system. Special attention to the scale betrays that the particles and wire widths in image **b** are much smaller than in **a**

illustrated above in Fig. 7.15. With this type of deposition, it was finally possible to create contact pads and electrical leads without altering or destroying the nanoline structures. A comparison of the resulting nanowire arrays with both deposition methods is shown in Fig. 7.16.

Nanowire Enhancement

From Fig. 7.16 we observe that the electrostatic deposition method results in nanolines in which the interparticle gaps are much smaller. Despite this high particle density, the lines do not conduct and, therefore, are not yet nanowires. Incubation of the nanoline arrays in commercial enhancement solutions for 5 to 15 min increases

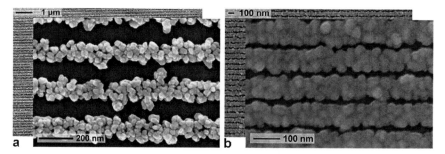

Fig. 7.17 Perspective SEM images of enhanced nanolines (nanowires) created by the self-assembly of DNA-tagged gold nanoparticles. **a** Enhanced 20-nm particles tagged with DNA and bound weakly via hybridization, approx. 130-nm line widths with a 200-nm period (*line/background*). **b** Enhanced 5-nm particles tagged with DNA bound strongly via electrostatic interaction, approx. 80-nm line widths with a 100-nm period (*line/background*). Special attention to the scale betrays that the particles and wire widths in image **b** are much smaller than in **a**

the size of the individual particles. The particles, therefore, grow toward each other, closing the interparticle gaps and resulting in conductive nanowires. Figure 7.17 shows nanowire arrays after such an enhancement process.

Not only are conductive nanowire arrays the result, but such enhancement processes also enable a controllable method for determining the interparticle gap size and the space between the parallel nanowires. Therefore, a combination of EUVL and nanowire enhancement allows us to tailor the overall density, and thus the optical, electrical, and mechanical properties, of the nanowire array.

Microfabrication and Micro-/Nanointerfaces

The electrostatically deposited DNA-tagged gold particles bind strongly to the surface. The particles, as well as the patterned structure (e.g., nanowires), survive sonication, additional photoresist exposure and development, as well as plasma treatments. Conventional lithographic techniques are now feasible for the creation of contact pads and electrical leads, which are critical for the reliable characterization of nanowire arrays. Figure 7.18 shows contact pads placed on top of and at the edges of a nanowire array.

Electrical Characterization

In order to determine the performance and the range of application for nanowire arrays, electrical characterization measurements are necessary. Metal nanowires, created with standard deposition methods, serve mainly as a reference for particle-based wire systems. Metal nanowires should have lower individual resistances than

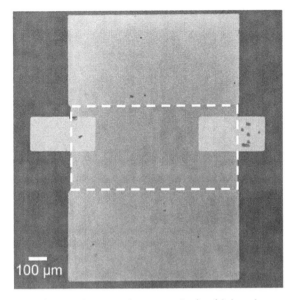

Fig. 7.18 SEM image of conductive nanowire array and microfabricated contact pads. The high-lighted center region is the nanowire array $(0.5 \times 1 \text{ mm}^2)$, and this corresponds to the interference pattern of EUV-IL. The regions above and below the nanowire array are nanoparticle films, emerging after total exposure (i.e., no interference pattern) during EUV-IL. The contact pads on the left and right edges of the nanowire array are solid metal structures, approx. 5 nm Ti and 50 nm Au. The observable scratches on the contact pads and the nanoparticle films mark the contact points from needles used for electrical characterization of the various regions of the sample

particle-based nanowires and, therefore, are more similar to bulk systems. However, because both metal nanowires and particle-based nanowires are created from the same EUV-IL nanoline patterns, they have the same dimensions and size scales.

More sophisticated samples and measurement systems for the in situ electrical characterization of the following nanowire arrays are currently being developed. Thus all results presented in this section have been obtained under dry conditions.

Metal Nanowire Arrays

Four-point and two-point measurements were performed on metal nanowire arrays with varying wire widths to determine the resistance of the nanowire array. Figure 7.19 shows such an array with contact pads of $300 \times 300 \, \mu\text{m}^2$. Since the wires have a 100-nm period, this results in a maximum of 3,000 wires in contact with the pads. This assumes, however, that all wires are continuous with no failures. This is seldom the case, because any interference in the lithographic process (e.g., spinning errors due to dust, scratches in the photoresist, etc.) and lift-off process could destroy certain wire regions.

The measureable arrays had widths near 50 nm and a deposited gold metal layer 50 nm in height. As represented in Fig. 7.20, the measured resistances ranged

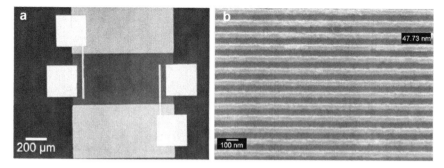

Fig. 7.19 a Optical microscope image of a metal nanowire array with contact pads arranged to allow various two-point and four-point measurements. **b** SEM image of deposited metal nanowires from **a**

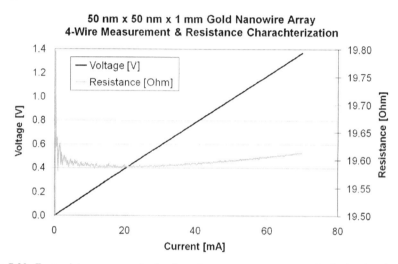

Fig. 7.20 Four-point measurement of voltage in response to current and calculated resistance of nanowire array. Assuming perfectly continuous nanowires with 3,000 parallel contacts to the measurement pads, a per-wire resistance estimate of 60 kΩ is obtained

between 17 and 20 Ω, resulting in an estimated resistance of 50 to 60 kΩ per nanowire. At currents above 30 mA, the resistance slowly rose in response to temperature effects and the heating of the wire.

Particle-Based Nanowire Arrays

Particle-based nanowires are expected to have a higher resistance mostly due to the increased scattering caused by the higher number of grain boundaries. Even in discontinuous wires, where the largest interparticle gap is less than 1 nm, tunneling is possible, but the resistance increases exponentially with the size of the gap. Any

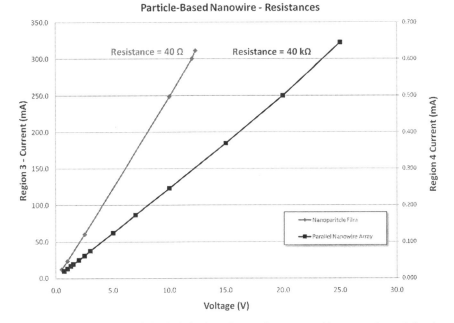

Fig. 7.21 Measurements of ohmic behavior of nanowire arrays, either overenhanced for the formation of a nanoparticle film (40 Ω) or enhanced enough to ensure conduction in parallel particle-based nanowires (40 kΩ). The size of the particles, the material, the width, the length, and the number of nanowires all change the conductive characteristics of the nanowire array, represented as a different I–V slope

interparticle gaps larger than 1 nm will inhibit tunneling currents, rendering the wire nonconductive over the length of the array.

In Fig. 7.21 initial results are shown in the form of a current–voltage measurement of an enhanced nanowire array. When a nanowire array is enhanced to the point where the parallel lines grow into each other, the result is similar to a particle film. Such an "overenhancement" has been shown to have a resistance of 40 Ω between the contact pads (Fig. 7.18). Under the same enhancement conditions, but in an adjacent nanowire array, where the separation between parallel lines was initially larger, overenhancement was avoided. Here, the electrons are confined to the individual wires of the array resulting in a higher resistance of 40 kΩ.

For a contact pad of 200 μm and a nanowire period of 100 nm, the maximum number of parallel connections is 2,000. This results in a minimum individual resistance of 80 MΩ per nanowire. One known gold-particle-based nanowire resistance measurement is from Blech et al. [45]. A reliable comparison, however, cannot yet be made since their nanowires are many orders of magnitude shorter and their approach to bridging the interparticle gap does not eliminate tunneling currents as well as the chemical enhancement of the gold particles.

The fabrication processes for both the nanowire arrays and the interfaces required to measure the nanowire arrays are not trivial. With increased knowledge and experience in the fabrication of nanowire-array-based electrical devices, our ability to investigate the dependent variables of conductance will improve. Knowing the real-world variables, the ultimate goal is then to construct nanowire-array-based systems with specific electrical properties (i.e., specific materials, specific gap sizes, built-in Coulomb blockages, etc.).

For biosensing applications it will be necessary to exploit those variables that maximize the local change in conductance ($\Delta G/G$). Alternatively, it might be possible to use the conductive nature of the nanowire arrays to influence or accelerate changes in electrochemical systems while simultaneously monitoring the systems optically. For this, it is necessary to also study the optical characteristics of the nanowire arrays.

Optical Characterization

Regular arrays of fine parallel metallic wires are known to function as absorptive polarizers, where a beam of electromagnetic waves is converted into a beam with a single polarization state. Therefore, it should be possible to observe polarization with our metallic nanowire arrays and particle-based nanowire arrays. Figure 7.22 confirms this by showing the polarization dependent spectra of gold-particle-based nanowire arrays.

By monitoring such spectra as shown in Fig. 7.22, it is also possible to, e.g., track the signal peak upon changes to the nanowire array. For this, the particle-based nanowires are particularly interesting, since their irregular form and edges could result in strong LSPR. Any biological analyte adsorption near or in these strong fields will have a strong influence on the extinction spectrum, which can be seen as a peak shift.

Optical Biosensing Measurements

By capitalizing on the nanogaps in the particle-based nanowire array, it was possible to construct a new biosensing platform with gold nanowires. With LSPR it was possible to observe changes in the extinction spectrum as biomolecules adsorbed to the surface of the exposed gold. Figure 7.23 shows the initial measurements using this technique with gold nanowire arrays. Here the change in the extinction spectrum is shown as a peak shift, or change in wavelength on the order of approx. 1.5 nm.

The nanoline arrays in the above measurements were composed of 5-nm, chemically enhanced particles. It would be possible to dramatically increase the signal-to-noise ratio of the system by constructing the nanoline arrays of larger gold particles, e.g., 10- or 20-nm particles. The larger particles yield an overall larger cross-sectional area of the wire, resulting in a stronger influence of the absorption and scattering (i.e., extinction) of the incident light spectrum. This, coupled with improvements to the spectrometer and measurement apparatus, as well as additional

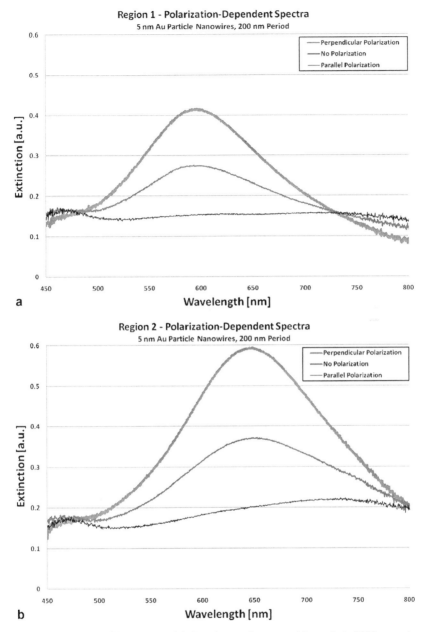

Fig. 7.22 Polarization of 5-nm Au particle-based nanowire array with a period of 200 nm. **a** Array of 140-nm-wide nanowires with a 60-nm separation between the wires. The result is the complete absorption of the electromagnetic energy of equally polarized light, while allowing a clear single-peak spectrum of light to pass through the transparent background when the incident beam is perpendicularly polarized. **b** Array of 110-nm-wide nanowires with 90-nm separation. The thinner wires absorb the electromagnetic spectrum slightly less effectively with parallel polarization, while the larger transparent background allows a comparatively stronger perpendicular polarization to pass through the wire array

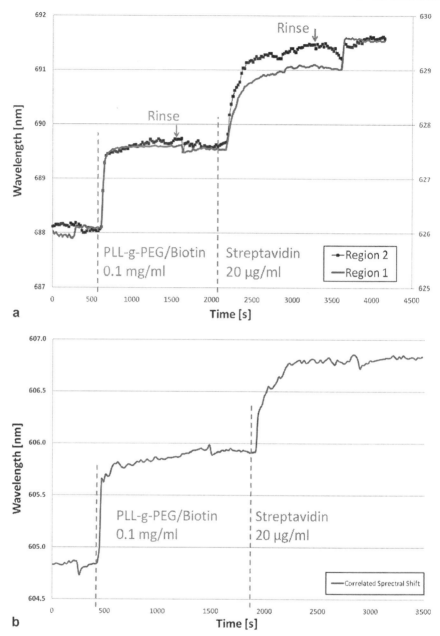

Fig. 7.23 Initial measurement using a particle-based nanowire array with a line period of 200 nm
(**a**) or 100 nm (**b**), produced with 5-nm enhanced Au particles. Based on LSPR and using a spectrometer, one can monitor the peak shift of the plasmon spectrum as molecules adsorb to the gold-particle surface, between the nanoparticles and nanolines. Our initial measurements were able to obtain a detection limit of 1 μg/ml of Streptavidin binding to adsorbed PLL–*g*-PEG–biotin

signal processing methods, is predicted to result in a limit of detection many orders of magnitude lower than the current 3σ value of approx. 1 µg/ml.

Summary

We presented a unique and emerging technology for the production of nanowire arrays. EUV-IL combines the advantages of large-scale pattern fabrication with high resolution (i.e., sub-10 m). In addition, the successful incorporation of nano-/ microinterconnects and the powerful characterization methods enabled a deeper understanding of the mechanical, electrical, and optical properties of the created functional nanowire arrays. This then allowed the realization of a new biosensing platform exploiting the phenomena of LSPR in nanowire arrays.

References

1. Natelson D (2003) Fabrication of metal nanowires. In: Recent developments in vacuum science and technology. Research Signpost, Trivandrum, pp 157–183
2. Cumming DRS et al (1996) Fabrication of 3 nm wires using 100 keV electron beam lithography and poly(methyl methacrylate) resist. Appl Phys Lett 68(3):322–324
3. Melosh NA et al (2003) Ultrahigh-density nanowire lattices and circuits. Science 300(5616): 112–115
4. Chai J et al (2007) Assembly of aligned linear metallic patterns on silicon. Nat Nanotechnol 2(8):500–506
5. Banqiu W, Kumar A (2007) Extreme ultraviolet lithography: a review. J Vac Sci Technol B Microelectron Nanometer Struct 25(6):1743–1761
6. Gonsalves KE et al (2005) High performance resist for EUV lithography. Microelectron Eng 77(1):27–35
7. Petrillo K et al (2007) Are extreme ultraviolet resists ready for the 32 nm node? 51st International Conference on Electron, Ion, and Photon Beam Technology and Nanofabrication. A V S American Institute of Physics, Denver, CO
8. Solak HH (2006) Nanolithography with coherent extreme ultraviolet light. J Phys D Appl Phys 39(10):R171–R188
9. Solak HH et al (2007) Photon-beam lithography reaches 12.5 nm half-pitch resolution. J Vac Sci Technol B 25(1):91–95
10. Auzelyte V et al (2007) Large area arrays of metal nanowires. 33rd International Conference on Micro- and Nano-Engineering. Elsevier, Copenhagen, Denmark
11. Olliges S et al (2009) Thermomechanical properties of flexible substrate supported gold nanowires. Scr Mater 60(5):5–8
12. Olliges S et al (2007) Tensile strength of gold nanointerconnects without the influence of strain gradients. Acta Mater 55(15):5201–5210
13. Olliges S et al (2008) In-situ observation of cracks in gold nano-interconnects on flexible substrates. Scr Mater 58(3):175–178
14. Bohm J et al (2004) Tensile testing of ultrathin polycrystalline films: a synchrotron-based technique. Rev Sci Instrum 75(4):1110–1119
15. Eberl C et al (2004) Ultra high-cycle fatigue in pure Al thin films and line structures. Symposium on Internal Stress and Thermo-Mechanical Behavior in Multi-Component Materials Systems held at the TMS Annual Meeting. Elsevier, Charlotte, NC

16. Patterson BD et al (2005) The materials science beamline at the swiss light source: design and realization. Nucl Instrum Methods Phys Res A 540(1):42–67
17. Schmitt B et al (2001) Mythen detector system. 10th International Workshop on Vertex Detectors. Elsevier, Brunnen, Switzerland
18. Schmitt B et al (2003) Development of single photon counting detectors at the Swiss Light Source. 9th Pisa Meeting on Advanced Detectors. Elsevier, La Biodola, Italy
19. Xia ZC, Hutchinson JW (2000) Crack patterns in thin films. J Mech Phys Solids 48(6–7): 1107–1131
20. Bohm J et al (2004) A new synchrotron-based technique for measuring stresses in ultrathin metallic films. Symposium on Nanoscale Materials and Modeling held at the 2004 MRS Spring Meeting. Materials Research Society, San Francisco, CA
21. Wu B, Heidelberg A, Boland JJ (2005) Mechanical properties of ultrahigh-strength gold nanowires. Nat Mater 4(7):525–529
22. Noyan IC, Cohen JB (eds) (1987) Residual stress. Springer, New York
23. Xiang Y, Vlassak JJ (2005) Bauschinger effect in thin metal films. Scr Mater 53(2):177–182
24. Xiang Y, Vlassak JJ (2006) Bauschinger and size effects in thin-film plasticity. Acta Mater 54(20):5449–5460
25. Dundurs J, Bogy DB (1968) Edge-bonded dissimilar orthogonal elastic wedges under normal and shear loading. J Appl Mech 35:460–466
26. Beuth JL (1992) Cracking of thin bonded films in residual tension. Int J Solids Struct 29(13):1657–1675
27. Gruber P et al (2004) Size effect on crack formation in Cu/Ta and Ta/Cu/Ta thin film systems. Symposium on Nanoscale Materials and Modeling held at the 2004 MRS Spring Meeting. Materials Research Society, San Francisco, CA
28. Fleck NA et al (1994) Strain gradient plasticity – theory and experiment. Acta Metallurgica Et Materialia 42(2):475–487
29. Gao H et al (1999) Mechanism-based strain gradient plasticity – I. Theory. J Mech Phys Solids 47(6):1239–1263
30. Gryaznov VG, Trusov LI (1993) Size effects in micromechanics of nanocrystals. Prog Mater Sci 37(4):289–401
31. Shen YF et al (2005) Tensile properties of copper with nano-scale twins. Scr Mater 52(10): 989–994
32. Courtney T (2000) Mechanical behavior of materials, 2nd edn. McGraw-Hill, New York
33. Gruber PA et al (2008) Temperature dependence of mechanical properties in ultrathin Au films with and without passivation. J Mater Res 23:2406–2419
34. Arzt E et al (1996) Physical metallurgy of electromigration: failure mechanisms in miniaturized conductor lines. Zeitschrift für Metallkunde 87(11):934–942
35. Agrawal PM, Rice BM, Thompson DL (2002) Predicting trends in rate parameters for self-diffusion on FCC metal surfaces. Surf Sci 515(1):21–35
36. MacKenzie R et al (2008) Electrochemical biosensors – Sensor principles and architectures. Sensors 8(3):1400–1458
37. Khanal DR et al (2007) Effects of quantum confinement on the doping limit of semiconductor nanowires. Nano Lett 7:1186–1190
38. Foley EL et al (1999) An undergraduate laboratory experiment on quantized conductance in nanocontacts. Am J Phys 67(5):389–393
39. Stadler B et al (2007) Nanopatterning of gold colloids for label-free biosensing. Nanotechnology 18(15):6
40. Maheshwari V, Kane JR, Saraf F (2008) Self-assembly of a micrometers-long one-dimensional network of cemented au nanoparticles. Adv Mater 20:284–287
41. Sannomiya T, Hafner C, Voros J (2008) In situ sensing of single binding events by localized surface plasmon resonance. Nano Lett 8(10):3450–3455
42. Stadler BM et al (2006) Light-induced in situ patterning of DNA-tagged biomolecules and nanoparticles. IEEE Trans Nanobiosci 5(3):215–219
43. Yu AA et al (2005) Supramolecular nanostamping: Using DNA as movable type. Nano Lett 5(6):1061–1064

44. Falconnet D et al (2004) A combined photolithographic and molecular-assembly approach to produce functional micropatterns for applications in the biosciences. Adv Funct Mater 14(8):749–756

45. Blech K et al (2008) In-situ electrical addressing of one-dimensional gold nanoparticle assemblies. J Nanosci Nanotechnol 8(1):461–465

Index

Printed in the United States
150944LV00003B/90/P